# Unless Recalled Earlier

**DATE DUE**

| | |
|---|---|
| | |
| | |
| | |
| | |
| | |
| | |
| | |
| | |
| | |
| | |
| | |
| | |
| | |
| | |
| | |
| | |

DEMCO, INC. 38-2931

# Dynamic and Stochastic Efficiency Analysis

## Economics of Data Envelopment Analysis

# Dynamic and Stochastic Efficiency Analysis

## Economics of Data Envelopment Analysis

**Jati K. Sengupta**
*University of California, Santa Barbara*

**World Scientific**
*Singapore • New Jersey • London • Hong Kong*

*Published by*

World Scientific Publishing Co. Pte. Ltd.

P O Box 128, Farrer Road, Singapore 912805

*USA office:* Suite 1B, 1060 Main Street, River Edge, NJ 07661

*UK office:* 57 Shelton Street, Covent Garden, London WC2H 9HE

**British Library Cataloguing-in-Publication Data**
A catalogue record for this book is available from the British Library.

DYNAMIC AND STOCHASTIC EFFICIENCY ANALYSIS: ECONOMICS OF DATA
ENVELOPMENT ANALYSIS

ISBN 981-02-4266-2

This book is printed on acid-free paper.

Printed in Singapore by Regal Press (S) Pte. Ltd.

To

**Mother**

With devotion

# Foreword

Measuring the economic performance of firms and industries through productivity studies has been a central concern for the economist and the operations researcher alike. This monograph develops and extends the recent methods of dynamic and stochastic analysis of economic efficiency by integrating two approaches: one is data envelopment analysis (DEA) developed in operations research theory and the other is Pareto efficiency applied in economic theory under competitive equilibrium.

The interface between economic theory and operations research is highlighted here through the analysis of market demand and optimization to achieve allocative efficiency. The DEA applications here emphasize the set of strategies available for the private sector firms facing competitive markets. Unlike the development of DEA theory in the last decade mostly restricted to nonprofit public sector organizations, our approach here stresses the impact of market competition and demand uncertainty on the cost and profit frontier of private firms.

Some of the outstanding features of this monograph are: (1) the use of learning by doing techniques in DEA models, (2) a two-stage analysis of market demand and production efficiency, (3) the conomics of general equilibrium in DEA theory, (4) the analysis of efficiency dynamics in terms of productivity growth and (5) the stochastic aspects of DEA efficiency in terms of the stochastic dominance approach.

The empirical applications here discuss the real life examples of various industries such as international airlines, electric power, telephones and the mutual fund industry. These applications illustrate the use of new statistical tools and nonparametric techniques to answer various managerial questions and policy issues. This monograph stresses the various applied aspects of dynamic and stochastic efficiency theory, so that the DEA model can be empirically implemented in diverse economic situations.

This valume includes a large part of my research work on DEA theory over the last five years. My sincere appreciation to all the research students who have helped me develop my views and empirically implement them.

Finally, I must record deep appreciation to my wife, who has provided constant support and encouragement. As always this work would not have been completed without her abiding support.

Jati K. Sengupta

# Contents

# 1. New Efficiency Theory

*University of California, Santa Barbara*

New efficiency theory known as 'data envelopment analysis' (DEA) purports to measure the relative effiiency of a set of units that are similar. The units are often called 'decision making units' or DMUs and their similarity lies in their pattern of input and output processes. When data are available for input costs and revenues from sale of output, the DEA model can be applied to firms in an industry to analyze their *technical* (production) and *allocative* efficiency. Technical efficiency measures the DMU's success in producing the maximum possible output from a given set of inputs, while allocative efficiency measures the firm's success in choosing an optimal set of inputs with a given set of input and/or output prices Clearly the allocative efficiency concept is most suitable for profit-oriented firms facing competitive markets.

The development of new efficiency theory based on the DEA approach has followed three important phases. The first phase started with the engineering concept of efficiency as a ratio of weighted outputs to weighted inputs and this concept of performance efficiency was applied by Charnes, Cooper and Rhodes (1978) in a linear proramming (LP) formulation to compare the relative efficiency of a set of DMUs. A similar approach was developed by Farrell (1957) to compare the relative efficiency of a cross-section sample of agricultural farms, but this was limited to one output for each firm. Note that this approach required the empirical data on the input and output quantities only and no prices are required. Hence only technical efficiency can be estimated by this approach.

The second phase introduced the concept of allocative efficiency, which leads to the specification of a cost frontier instead of a production frontier. The econometric studies of cost frontier functions have frequently used the allocative efficiency criterion based on the observed price data.

The third phase extended the cost efficiency approach further, by using inputs and/or outputs as policy variables to be optimally chosen by each firm or DMU, when it faces market prices under perfect or imperfect competition. Two features of flexibility in this approach are to be noted. One is that the optimal input levels can be used for future by firms, who have to adjust their inputs accordingly. This permits a dynamic view of the optimal input paths over time, when market prices change in a predictable manner. Secondly, the stochastic aspects of learning and adjustment, when firms adjust their inoptimal input levels to the optimum, can be directly introduced in this framework.


## 1.1 Economics of Data Envelopment Analysis

The economists' view of the DEA approach is that it is a technical method of efficiency measurement that is unrelated to the economic environment under which firms or DMUs operate.

  For public sector applications the DEA approach basically computes a set of optimal weights $\beta^*$ and $\alpha^*$ say for the input (x) and output vectors (y) and measures efficiency for DMU$_j$ by the ratio $\alpha^*{}' y_j / \beta^*{}' x_j$ for each j=1,2,...,N. By the convexity of the producton set we have $\alpha^*{}' y_j \leq \beta^*{}' x_j$, when prime denotes the transpose of a vector. Hence if DMU$_j$ is technically efficient we have

$$\alpha^*{}' y_j = \beta^*{}' x_j \; ; \alpha^*, \beta^* \geq 0$$

Two points are to be noted. First, the optimal weights $(\alpha^*, \beta^*)$ are specific to each firm or DMU. Hence there exist in principle N pairs of optimal weights, all of which may not be distinct. This makes it difficult to interpret $\alpha^*, \beta^*$ as the shadow prices of outputs and inputs or, as maket prices in some sense. Secondly, the cost aspect of the input utilization process is completely ignored. Even for public sector enterprises the costs of inputs are largely market determined but the DEA models fail to introduce any theory of market competition. Indivisibility of some of the inputs and also the productivity due to research inputs and learning by doing, which are so strongly emphasized in economic theory are rarely discussed in the DEA literature.

  In private sector applications to firms in an industry the DEA models have to compete with the economic models based on profit maximization under perfect or imperfect market competition. Under competitive markets when the firms are price takers, the equilibrium requires two conditions: (i) the firm has to choose optimal inputs and outputs that maximize profits, and (ii) the total industry demand whould determine the number N of firms which would survive. The first is the individual profit maximization objective for each firm, while the second is based on the assumption of free entry and free exit. The DEA model makes no distinction between equilibrium and disequilibrium points in the input output space and the industry effect on the individual firms is totally ignored.

  In the recent DEA literature two types of comparisons with the economic models are usually attempted. One is to compare the DEA measures of technical efficiency with suitable measures of profits. Since firms operate in different environments resulting in unequal profit contributions, their overall profitability depends very critically on the different environmental factors which are outside of the control of the individual firms. The DEA model does not adequately incorporate these extraneous factors and hence it happens more often that firms which are technically efficient by the DEA approach are not always the highest in profitability. Thus Schefczyk (1993) found in his DEA application to international airlines data of 14 airlines over the period 1988-90 that gross profit margins do not explain more than 32% of their technical efficiency.

A second approach is to separate the two stages in the firm's decision making process. The first stage, called market efficiency relates to the firm attracting demand and generating revenue. Advertising expenditure and marketing inputs are utilized in this stage to optimally influence overall demand. The second stage minimizes total operating costs for a given level of demand and it is termed cost efficiency. The two stages are linked through the market demand and the assumptions of free entry and free exit can be easily introduced through their effects on the prices. Athanassopoulos and Thanassoulis (1995) have applied this two-stage form of the DEA model to analyze the performance of a chain of grocery stores facing spatial competition through demand and costs.

## 1.2 Dynamic and Stochastic Efficiency

The sources of dynamics in new efficiency theory are several. On the input side the durability of inputs such as capital stocks and technical change either embodied in specific inputs or disembodied as experience and skill, is the major source of dynamics. The rates of utilization of the durable inputs in the short run do not reflect the long run characteristics of the output profile over time. Any static DEA model which ignores the output profile over time is likely to be seriously biased. Technology has a similar impact on the output path over time. A second important source of dynamics is the phenomenon of 'learning by doing'. Cumulative experience gained through learning affects productivity growth and it may affect different firms differently. These learning curve effcts have played a significant role in modern technology-intensive industries such as microelectronics, semiconductors and software development.

A third source of dynamics arises when firms attempt to choose optimal inputs and outputs over time in order to achieve allocative efficiency. Since future prices and demand are not always known with perfect certainty, any optimal plan becomes partly stochastic in character. In such a framework economic behavior frequently involves lags in adjusting stocks of inputs to their desired levels. For instance, a firm or DMU which finds that its current input stocks are inconsistent with the long run optimal path implied by current relative factor prices will geneally spread the planned adjustment to long run optimal levels over a period of time. Thus the expected costs of adjustment and disequilibrium can be minimized in a dynamic DEA model of allocative efficiency.

Finally, dynamics in a growth context involves the issue of output growth over time and the extent to which it can be explained by the input growth of different inputs and their marginal contributions. Solow-type growth models in macrodynamics have adopted this approach and numerous empirical applications over international cross-section and time series data have been reported in recent economic literature. Clearly this approach can be applied in the DEA framework by replacing the input and output levels by the growth in inputs and outputs, with some inputs being capital and technology.

A major criticism of the DEA model of efficiency is that it ignores the fact that the input output data are not all deterministic due to measurement errors and disturbances. Stochastic frontier analysis introduces two types of disturbances or errors, one being a symmetric component with zero mean and the other a positive disturbance. Thus in the single output case of a linear model we have for firm j:

$$y_j = y_j^* - u_j; u_j \geq 0; j = 1, 2, \ldots N$$

$$y_j^* = \sum_{i=0}^{m} \beta_i x_{ij} + e_j; E(e_j) = 0$$

where $y_j^*$ is the maximal output, $x_{ij}$ are the inputs and $u_j$ denotes efficiency differences across firms that are random. The other random component $e_j$ is the symmetric error with a zero mean.

Two other aspects of stochasticity are important in the DEA approach to efficiency. One is the attitude to risk adopted by each agent or DMU. For a risk averse agent the response to input output fluctuations would be one of hedging against uncertainty and this would be quite different from the attitude of a risk taking agent. The transformation of the DEA model needed to allow for these risk averse or risk taking attitudes provides an important avenue of extension of the DEA approach. An important area of application of this transformation approach is in comparing the relative performance of investment in different mutual funds. Since the investment market is risky, different types of mutual fund e.g., growth fund, balanced fund and income fund provide different degrees of risk aversion. If each mutual fund is considered a DMU, the various inputs are the expense ratio, the turnover costs and standard deviation of return, while the outputs are capital gains and dividends which may be combined as mean return.

Another stochastic feature of the optimal solutions of a DEA model is the type of statistical distribution to which it leads. If the efficiency scores arising from technical efficiency measurement in DEA models are used to compute the efficiency distribution, then its characteristics could be utilized to estimate the mean, variance and other higher moments. It can also provide some information for applying maximum likelihood methods of estimation in stochastic frontier analysis. For example if the efficiency distribution follows a gamma distribution, its parameters can be separately estimated by maximum likelihood methods. Thus one could combine the DEA method with the maximum likelihood method of stochastic frontier anlaysis.

In case of allocative efficiency when inputs are optimally chosen, the stochastic fluctuations in input prices or costs would affect the distribution of optimal inputs and output and once again the various moments e.g., mean, variance, skewness of the distribution may be computed. Clearly these moments would characterize the various types of market fluctuations in input prices.

## 1.3 Cost Frontier Anlaysis

By duality theorm a cost frontier can be derived from a production frontier, if the market prices are given. Even in cases of increasing returns to scale to production, when a profit function becomes unbounded, a cost function is well defined and a firm can minimize overall costs for any given level of output. Thus one can derive a cost frontier as follows

$$\text{Min } C = \Sigma \; q_i \; x_i$$
$$\text{subject to } y \leq f(x_1, x_2)$$

where $q_i$ are the input prices or unit costs and $f(x_1, x_2)$ is the production funciton. If the production function is of a Cobb-Douglas form $f(x_1, x_2) = A \; x_1^{b_1} x_2^{b_2}$, the cost minimization model above yields the cost frontier

$$\ell n \; c^* = k_0 + \sum_{i=1}^{2} \; k_i \ell n \; q_i + a \; \ell n \; y *$$

where $k_i = b_i / \Sigma \; b_i$, $a = (b_1 + b_2)^{-1}$ and $k_0$ is a constant intercept depending on $b_1$, $b_2$ and A. The asterisks denote optimal values. Clearly any observed cost $c_j$ for firm j can be written as

$$\ell n \; c_j = \ell n \; c_j^* + v_j; v_j \geq 0$$

In order to test the relative cost efficiency of a unit k, a DEA model may then be set up as follows:

$$\text{Min } v_k = \text{Max } \ell n \; c_k^*$$
$$\text{s.t. } \ell n \; c_j^* \leq \ell n \; c_j; j = 1, 2, ..., N$$

If unit k is cost efficient, then $\ell n \; c_k^* = \ell n \; c_k$; otherwise it is relatively inefficient.

This type of cost minimization model is equally relevant for a nonprofit organization as for a profit-oriented firm.

Consider a linear model where the inputs $x_{ij}$ are replaced by their costs $C_{ij}$ and outputs $y_{rj}$ by their returns $R_{rj}$. Then we setup the following DEA model in order to test the relative efficiency of a unit k:

$$\text{Min } \theta$$
$$\text{s.t } \sum_{j=1}^{N} C_j \lambda_j \leq \theta C_k ; \sum_{j=1}^{N} R_j \lambda_j \geq R_k \qquad (1.1)$$
$$\sum \lambda_j = 1; \lambda_j \geq 0; \quad j = 1, 2, ..., N$$

where $C_j$, $R_j$ are the cost and revenue vectors for unit j and the decision variables are $\lambda_j$ and $\theta$. Here $\theta$ is a radial measure of efficiency i.e., efficiency score defined in terms of costs. Once again if $\theta^* = 1.0$ and $\sum_j C_j \lambda_j^* = C_k$ with $\sum R_j \lambda_j^* = R_k$ we have $DMU_k$ cost efficient. In case we allow for both input-specific and output-specific efficiency components, then the model can be transformed as:

$$\text{Min } \sum_{i=1}^{m} \theta_i - \sum_{r=1}^{s} u_r$$
$$\text{s.t. } \sum_j C_{ij} \lambda_j \leq \theta_i C_{ik} ; \sum_j R_{rj} \lambda_j \geq u_r R_{rk}$$
$$\sum \lambda_j = 1; \lambda_j > 0; j = 1, 2, ..., N$$

If the output prices are not market determined, we have to replace the revenue vector $R_j$ by the output vector $Y_j$. Furthermore if $DMU_k$ is cost efficient in DEA model (1) with $\theta^* = 1.0$ then the optimal profit $\pi_k^*$ is zero, where

$$\pi_k^* = \alpha^{*\prime} R_k - \beta^{*\prime} C_k - \beta_0^*$$

and the row vectors $\alpha^{*\prime}, \beta^{*\prime}$ are optimal Lagrange multipliers.

An alternative way to view the cost function is to relate costs to various otuputs and then set up a cost minimization model. For example consider aggregate cost for each $DMU_j$ as a linear function of s outputs and $v_r$ as the marginal cost of output $y_r$. Then to test the cost efficiency of $DMU_k$ one sets up the following model:

$$\text{Max } \sum_{r=1}^{s} v_r y_{rk}$$
$$\text{s.t. } \sum_{r=1}^{s} v_r y_{rj} \leq C_j ; j = 1, 2, ..., N \qquad (1.2)$$
$$\sum_r v_r = 1, v_r \geq 0$$

Here $C_j$ is the total observed cost for $DMU_j$ and $\Sigma\ v_r = 1$ is the normalization condition for the cost components or shares. If $DMU_k$ is cost efficient then we must have

$$\alpha_0^* + \sum_{j=1}^{N} \alpha_j^* y_{rj} = y_{rk}$$

where $\alpha_j^*, \alpha_0^*$ are the Lagrange multipliers with $\alpha_j^*$ nonnegative and $\alpha_0^*$ free in sign.

An interesting application of the cost function approach is in measuring economies of scope. Let $C_j(y_1, y_2, ..., y_s;\ q_1, q_2, ..., q_m) = C_j(y, q)$ be a nonlinear e.g. quadratic, loglinear or translog function for $DMU_j$, then economies of scope is defined by

$$\text{Scope} = [C_j(y_1, 0, ..., q) + C_j(0, y_2, 0, ..., q) + ...$$
$$+ C_j(0, 0, ..., 0;\ q) - C_j(y;q)]/C_j(y;q) \qquad (1.3)$$

A more generalized concept of quasi-scope has been introduced by Pulley and Braunstein (1992), which distinguishes between quasi-specialized and non-specialized outputs and introduces a parameter w to be the proportion of non-specialized outputs produced. Thus

$$\text{Quasi-scope} = [C_j(\{1-(s-1)w\}y_1,\ wy_2, ..., wy_s;\ q)$$
$$+ C_j(wy_1, \{1-(s-1)w\}y_2 + ... + wy_s;q)$$
$$+ ... + C_j(wy_1, wy_2, ..., \{1-(s-1)w\}y_s;q)$$
$$- C_j(y;q)]/C_j(y;q) \qquad (1.4)$$

When w=0 this reduces to the traditional scope measure. Clearly the quasi-scope measure can be introduced into the DEA model (2) by running a pair of nonlinear programs involving $C_j(y;q)$ and the cost function $[C_j(\{1-(s-1)w\}y_1,\ wy_2, ..., wy_s;q) + C_j(wy_1, \{1-(s-1)w\}y_2 + wy_3 + ..., q) + ... + C_j(wy_1, ..., \{1-(s-1)w\}\ y_s;q]$.

Note that for $w > 0$ quasi-scope economy is an empirical subadditivity measure capturing both scope and scale effects of a s-component division of total output. The maximum value for w is 1/s. here $\{1-(s-1)w\} = 1/s$ and the qausi-specialized production amounts to producing the proportion 1/s of all outputs. At this point joint production is no longer specialized.

One interesting economic application of the cost frontier approach is to incorporate the impact of R&D investment on unit costs. Consider the cost function of $DMU_j$ as

$$C_j = \sum_{r=1}^{s} \left( (\beta_r - a_r) y_{rj} + \gamma_r y_{rj}^2 \right)$$

where the positive value of $a_r$ indicates the impact of research and development expenditure. Initially $a_r$ is zero when R&D expenditure is low or negligible. The dependence of $a_r$ on R&D can be indicated by a function $a_r = a_r(R\&D)$. To test the cost efficiency of a unit k we set up the DEA model which is linear in parameters $\beta_r$, $a_r$ and $\gamma_r$:

$$\underset{\beta_r, \gamma_r, a_r}{\text{Max}} \sum_{r=1}^{s} \left[ (\beta_r - a_r) y_{rk} + \gamma_r y_{rk}^2 \right]$$

$$\text{s.t.} \sum_{r=1}^{s} \left[ (\beta_r - a_r) y_{rj} + \gamma_r y_{rj}^2 \right] \le C_{rj} \quad j=1,2,\ldots,N$$

$$\beta_r, a_r, \gamma_r; r = 1,2,\ldots,s$$

One can easily introduce an intercept parameter into the cost function and hence the dual form can be written as

$$\text{Max } \psi$$

$$\text{s.t.} \quad \sum_{j=1}^{N} y_{rj} \lambda_j \ge \psi y_{rk}; \quad r = 1,2,\ldots,s$$

$$\sum_{j=1}^{N} y_{rj}^2 \lambda_j \ge y_{rk}^2; \quad r = 1,2,\ldots,s$$

$$\sum \lambda_j = 1, \lambda_j \ge 0$$

Another interesting feature of the cost frontier approach in DEA framework is that component costs such as expected inventory costs due to supply exceeding demand or costs due to the adoption of flexible technologies as in FMS (flexible manufacturing systems) may be directly introduced. In the last case the advantages of adopting a flexible technology instead of a fixed one dedicated to specific outputs can be directly measured by means of a DEA model.

## 1.4 Market Demand and Efficiency

From an economist's viewpont the single most important weakness of the DEA model is that it ignores the demand characteristics of the input and output markets. In production or technical efficiency the observed inputs and outputs are the only data considered in developing the efficiency scores. But even for public sector units the inputs like labor and materials are bought in competitive markets and hence the type of market structure would influence the pattern of input use.

In private sector applications where profit is generally used as the criterion for determining the optimal input and output levels, we are presented with a choice problem: which measure of efficiency should we use? The DEA measure or the allocative efficiency measure or profits? Once again the market competition and the degree of imperfection if any have significant impact on the possible answer to the question. The influence of market demand on the DEA measure of allocative efficiency may be incorporated in several ways. One method is a two-stage approach. In the first stage the ojbective is to model the firms' behavior in attracting demand and generating revenue. Advertising and marketing inputs are optimally used to influence the market demand along with the pricing strategy, if the firms have some control on the prices as in monopolistic competition. The second stage minimizes total operating costs for a given revenue level and this measures the cost efficiency of a firm.

A second approach is to incorporate expected inventory costs as one of the cost components, where inventory costs include the impact of demand on the supply of various outputs. Thus assuming quadratic inventory costs and linear production costs, one could set up the following DEA model of profit maximization:

$$\text{Max} \pi = \sum_{r=1}^{s} p_r y_r - \sum_r w_r y_r - (1/2)\sum_r c_r (y_r - d_r)^2$$

$$\text{s.t.} \quad \sum_{j=1}^{N} y_{rj}\lambda_j \geq y_r; \quad r = 1,2,\ldots,s$$

$$\sum_j x_{ij}\lambda_j \leq x_{ik}; \quad i = 1,2,\ldots,m$$

$$\sum_j \lambda_j = 1, \lambda_j \geq 0; \quad j = 1,2,\ldots,N$$

Here the parameters $p_r$, $w_r$ and $c_r$ are given either by the market or by technical conditions of production. The decision variables are the various outputs ($y_r$) and the weight variable $\lambda_j$. Here the objective is to test the relative efficiency of unit k. If the unit k is efficient then its outputs $y_{rk}$ should equal the optimal solution $y_r^*$ of the quadratic program above. If the market prices $p_r$ are not known or inapplicable due to the nonprofit nature of the firms or DMUs, one minimizes overall costs given by

$$C = \sum_r w_r y_r + (1/2)\sum_r c_r (y_r - d_r)^2$$

this yields the optimal solution $y^* = (y_r^*)$ and $\lambda^* = (\lambda_j^*)$.

Note that the cost minimization model above can be applied in the input space also. For instance let $z_i$ be the demand for input i and $x_i$ is the input to be chosen optimally with given market prices $q_i$, then the model becomes:

$$\text{Min} \sum q_i x_i + (1/2)\sum_{i=1}^{m} h_i(x_i - z_i)^2$$

$$\text{s.t.} \quad \sum_{j=1}^{N} y_{rj}\lambda_j \geq y_{rk} \ ; r = 1,2,\ldots,s$$

$$\sum_{j=1}^{N} x_{ij}\lambda_j \leq x_i \ ; i = 1,2,\ldots,m$$

$$\sum \lambda_j = 1, \lambda_j \geq 0$$

If the observed value of $x_{ik}$ equals the optimal value $x_i^*$, then the $DMU_k$ has allocative efficiency. This suggests a two-stage interpretation of the allocative DEA models. Consider for example the scheduling model known as the HMMS model originally due to Holt, Modigliani, Muth and Simon (1960), that has been extensively applied in the management science literature. Following this model we define the following cost components:

$$C_{1t} \geq C_{1t}^* = c_1 x_{1t} \ \text{(regular payroll)}$$

$$C_2 \geq C_{2t}^* = c_2(x_{1t} - x_{1,t-1})^2 \ \text{(hiring and layoff)}$$

$$C_{3t} \geq C_{3t}^* = c_3(x_{2t} - c_4 x_{1t})^2 + c_5 x_{2t} - c_6 x_{1t} \ \text{(overtime and idle time)}$$

$$C_{4t} \geq C_{4t}^* = c_7(x_{3t} - c_8 - c_9 d_t)^2 \ \text{(inventory and shortages)}$$

where $C_{it}$ and $C_{it}^*$ are the observed and the minimal costs for different components ($i = 1,2,3,4$) and the inputs are: $x_1$ = volume of work force, $x_2$ = output, $x_3$ = inventory and $d_t$ is the estimated sales. Our samples are T periods ($t = 1,2,\ldots,T$) for which the data set $C_{it}$, $x_{jt}$ are observed. To test which of the sample points are technically efficient, we solve the nonlinear DEA model

$$\min_{c_1,\ldots,c_9} \sum_{i=1}^{4} (C_{ik} - C_{ik}^*) = \text{Max}_c \sum_{i=1}^{4} C_{ik}^*$$

$$\text{s.t.} \ C_{it} \geq C_{it}^* \ , i = 1,2,3,4; t = 1,2,\ldots,T$$

for the obsevation at the k-th time point ($k = 1,2,\ldots,T$). By varying k in the objective function over the sample set ($k = 1,2,\ldots,T$) one may test the DEA efficiency for each point. Let E be the set of DEA efficient observations out of the T sample points. Let $c^* = (c_1^*,\ldots,c_q^*)$ be the average (median) estimate from the set E. Given $c^*$ one may then set up at the second stage the dynamic decision model

$$\underset{x}{\text{Min}}\, C_\tau = \sum_{t=T+1}^{\tau} \sum_{i=1}^{4} C_{it}^*(x_1, x_2, x_3)$$

for determining the optimal levels of $x_{1t}$, $x_{2t}$, $x_{3t}$ over the future planning horizon t = T+1,…,τ. Note that this two-stage formulation is basically the same as before with two differences. One is that it is a nonlinear cost frontier, so that nonlinear programming methods have to be applied here in order to obtain the DEA parameters $c^* = (c_1^*, c_2^*, …, c_q^*)$. Secondly, the demand or sales variable $d_t$ has to be predicted, since there is an inventory constraint in the HMMS model e.g.,

$$x_{3,t} = x_{3,t-1} + x_{2t} - d_t \geq 0$$

with no backlogs permitted. Otherwise demand has to be modelled as a stochastic process and the objective function revised as the minimization of an expected cost function.

One important feature of this formulation is that it allows sequential updating of the optimal decision rule yielded by the second stage model as the time horizon τ is shifted forward over time.

Recently Kumar and Sinha (1998) compared the two methods of estimating the parameters of the cost function model of HMMS: the regression method and the DEA method. The first method estimated the average cost function, whereas the second method estimated the cost frontier by the DEA approach. The data used in the application were generated using a uniform distribution with means of various variables like demand and the various cost components approximately equal to those used by HMMS. The data on previous operations were generated for 36 consecutive time periods. In each of the 100 experimental cases it was observed that the total cost for the planning period is lower when production follows the DEA model than when using the regression method. The following table reports the results for four two data sets.

|  | DEA | | Regression | |
|---|---|---|---|---|
|  | Mean | Std. Dev | Mean | Std. Dev |
| Data set 1 | 497.5 | 18.2 | 945.9 | 35.3 |
| Data set 2 | 452.9 | 26.1 | 823.6 | 43.8 |

Table 1.4.1 Average Total Cost

In terms of sensitivity of the coefficient estimates the DEA method showed more robustness than the regression approach.

## 1.5 Managerial Perspectives

The application of DEA models in evaluating the production and allocative efficiency of a set of units raises several managerial issues. One of the most important practical problem is due to the fact that the input output data are generally noisy. Thus if the data can be decomposed into a systematic part and a random unsystematic part, the efficiency measurement and ranking by the DEA method should only be applied to the systematic part. This is where the quality control techniques appear to be most suitable for the DEA framework. The quality control theory assumes that the variations in the observed characteristics of outputs and sometimes the inputs is always due to a large variety of causes. Thus it would be almost impossible to enumerate the possible reasons why repeated outputs from a given input differ, or why successive batches of inputs fail to be identical. Control charts based on the concept of control limits are widely used in industry to detect assignable causes of variation in product quality which are nonrandom. The control limits most frequently used are three standard deviations of the average value. This generally covers 99.7% for the normally distributed data. If however some information is available about the specific distribution other than the normal e.g., a gamma density, the control limits can be appropriately modified. Clearly this technique would reduce the effect of outliers on the DEA estimates of systematic efficiency. A Bayesian technique of updating this method has been discussed by Sengupta (1996a, 1998) in case a panel data is available.

Another practical method for reducing the impact of random errors ($\varepsilon_{ij}$) in the input space is to introduce the cost of errors $c_{ij}$ ($\varepsilon_{ij}$) in the DEA model as:

$$\text{Min } \theta + \phi$$

$$\text{s.t.} \quad \sum_{j=1}^{N} x_{ij}\lambda_j \leq \theta\, x_{ik}$$

$$\sum_j c_{ij}(\varepsilon_{ij})\lambda_j \leq \phi c_{ik}(\varepsilon_{ik}); \ i = 1,2,\ldots,m$$

$$\sum_j y_{rj}\lambda_j \geq y_{rk}; \ r = 1,2,\ldots,s$$

$$\sum_j \lambda_j = 1, \lambda_j \geq 0; j = 1,2,\ldots,N$$

But since $c_{ij} = c_{ij}(\varepsilon_{ij})$ is not observable, we replace it by the standard deviation $\sigma_{ij}$ of input i for $DMU_j$. When panel data are available, estimates of $\sigma_{ij}$ are available and these can be used. Similarly the standard errors of the ouputs $y_{rj}$ can be incorporated.

Another important problem is due to the incentives needed to improve efficiency for the units which are relatively inefficient. Since the DEA model lacks information on each agent's production function, agent's effort on output is unknown and uncertain. Bogetoft (1994) has recently modelled this situation as an agency problem. For the one output case we let

$$y_j = (f(\underline{x}_j)a_j - \varepsilon_j$$

where $a_j$ is the agent's effort, $f(\underline{x}_j) > 0$ is the output per unit of effort and $\varepsilon_j$ is the agent-specific uncertainty. He assumes that the actual production frontier $f(\cdot)$ is unknown but the set of possible frontiers is known and the players' beliefs about the likelihood of different frontiers is given by a known probability distribution. Now we assume that the effort $a_j$ is decided by agent i and that it is nonobservable to the principal. What would the principal do in such a case? His goal is to minimize the expected payments needed to induce the agents to privately select the effort levels $a^* = (a_1^*, a_2^*, ..., a_N^*)$ which are optimal from the principal's viewpoint. But in this agent-principal framework we do not use agent j in the estimation of the frontier to which he himself is compared.

Finally, the allocative efficiency formulation of the DEA model makes it possible for the manager to compare his observed input and output levels with the optimal values and thereby adjust over time. The risk averse attitude of the manager to the input and output price fluctuations may also be incorporated here. The fact that some of these risk-oriented behavior and/or constraints may be implicitly known rather than explicitly observed makes it possible for different agents to have different degrees of absolute or relative risk aversion and this would seriously bias the relative efficiency measurement and ranking.

Some of the implicit constraints facing an agent may spill over to the other agents and this externality effect is very important in new industries such as microelectronics and telecommunications. For these resons the efficiency calculations in the DEA model may be second best optimal solutions rather than the first best.

# 2. Economics of Efficiency Measurement

The economists take a rather critical view of the DEA approach to efficiency measurement. Broadly speaking their criticisms can be summarized into three groups. First, the specification side of the approach does not pay adequate attention to the objectives or goals of the agents, their view of the markets for which they are producing and the problems of allocation of various inputs to the various outputs. True, it provides a method of ranking the different agents or DMUs in terms of the technical efficiency measure but it appears to be a mere technique of measurement without any solid economic theory. Second, the fact that the same set of inputs can lead to different levels of output due to random errors lead one to believe that the efficiency estimate by the DEA approach is not very robust. Not only the outliers but also the stochastic nature of the constraints may significantly affect the DEA efficiency estimate. Similarly the neglect of the dynamic dimensions of some inputs like capital and R&D inputs, which influence both current and future ouputs leads to the criticism that the DEA efficiency measures are myopic and static. Finally, the post-optimality veiw of the linear programming models used in the DEA approach appears to be weak, since it does not build into the approach any method of learning or adaptivity by those agents who are found to be less efficient at any given point of time. The structure of inefficiencies revealed by the subset of inefficient firms may need to be analyzed in more detail. The so-called profiling analysis in DEA approach e.g., Tofallis (1996) is an attempt in the right direction. In this analysis the first stage identifies the subset of efficient agents who are then further tested to see if some critical input e.g., capital is fully utilized or not. If it is not fully utilized, then an efficient agent may not be fully efficient in the utilization of the capital input.

This chapter attempts to answer some of these economic criticisms by providing the background of economic theory underlying the approach of data envelopment analysis.

## 2.1 Economics of the DEA Approach

We consider in this section three borad aspects of the economic framework underlying the DEA approach.

 A. A two-stage model
 B. Flexibility in production process
 C. Learning and cumulative experience

The two-stage model specifies the firm-specific decision problem first and then attempts an inter-firm comparison by the DEA approach. The flexibility analysis points out the importance of the optimal decision problem for a firm choosing between fixed and flexible technologies. Finally, the impact of learning through cumulative experience on the productive efficiency of firms is discussed in the DEA framework. All these formulations are intended to make explicit the various sources of productive efficiency underlying the input output data of firms in an industry.

## A. Two-stage Model

Consider the decision problem of a firm j, also called the decision making unit (DMUj) as a two-stage model. In the first stage the firm chooses the optimal input $x^*(j)$ and output $y^*(j)$ vectors by solving a standard linear programming (LP) model

$$\text{Max } z(j) = c'(j)\, y(j) - q'(j)\, x(j)$$
$$\text{subject to } A(j)\, y(j) \le x(j) \tag{2.1}$$
$$y(j) \ge 0$$

where the output and input prices $c(j)$ and $q(j)$ are given in the competitive market along with the input-output matrix $A(j)$ for each $j=1,2,\ldots,N$. Here prime denotes the tanspose of a vector. Let the optimal solution be $x^*(j)$, $y^*(j)$ and $z^*(j)$ for each firm j.

On using these optimal input and output vectors as data, the second stage sets up an inter-firm comparison model in order to measure the relative efficiency of a firm k where k is any firm in the index set $I_N = \{1,2,\ldots,N\}$. This stage uses a DEA approach to compute technical ($TE_k$) and allocative efficiency ($AE_k$) of $DMU_k$ relative to all other DMUs in the set. For technical efficiency measurement one sets up the LP model as:

$$\text{Min } \theta$$
$$\text{s.t. } \sum_{j=1}^{N} x^*(j)\lambda_j \le \theta x^*(k)$$
$$\sum_{j=1}^{N} y^*(j)\lambda_j \ge y^*(k) \tag{2.2}$$
$$\sum_{j} \lambda_j = 1,\ \lambda_j \ge 0$$

Here k belongs to the index set $I_N$ and for a fixed k, $\theta^* = \theta_k^*$ is the optimal value that is used to characterize technical efficiency of $DMU_k$. Let $\theta^* = \theta_k^*$ and

$\lambda^* = (\lambda_j^*)$ be the optimal solutions of the above DEA model (2.2) with all the slack variables zero. Then the reference unit k or $DMU_k$ is *technically efficient* if $\theta^* = 1$ and the first two sets of inequalities in (2.2) hold as equality in the optimal basis. If $\theta^*$ is positive but less than unity, then the $DMU_k$ is not technically efficient at the 100 percent level.

Overall efficiency ($OE_k$) of a $DMU_k$ or firm k however combines both technical ($TE_k$) or production efficiency and the allocative ($AE_k$) or price efficiency as follows:

$$OE_k = TE_k \cdot AE_k; \quad k \in I_N \tag{2.3}$$

To characterize overall efficiency of a firm or $DMU_k$ one sets up an LP model for minimizing total input costs $C = q'x$ as follows:

$$\text{Min } C = q'x$$
$$\text{s.t.} \quad \sum_{j=1}^{N} x^*(j)\lambda_j \leq x; \sum_{j=1}^{N} y^*(j)\lambda_j \geq y^*(k)$$
$$\sum_{j=1}^{N} \lambda_j = 1; \lambda_j \geq 0, j \in I_N \tag{2.4}$$

Here q is an m-element vector of input costs or input prices as determined in the competitive market and x is an m-element input vector to be optimally decided by $DMU_k$ along with the nonnegative weights $\lambda_j$. This is an input-oriented version of the DEA model for determining overall efficiency. Let $\lambda = (\lambda_j^*)$ and x* be the optimal solutions of the DEA model (2.4) with all slack variables zero in the optimal basis. Then the minimal input cost is given by $C^* = q'x^* = C_k^*$, whereas the observed cost is $C = q'x(k)$. Hence we obtain the three efficiency measures as follows:

$$OE_k = C_k^* / C_k; TE_k = \theta^* = \theta_k^* \text{ and } AE_k = OE_k / TE_k$$

Note that the iput vector x in (2.4) is a decision vector to be optimally chosen by $DMU_k$, whereas x(k) is the observed input vector in DEA model (2.2). If $\theta^*x(k) = x^*$, then $DMU_k$ is both technically and allocatively efficient, i.e., $TE_k = 1.0 = AE_k$ and hence $OE_k = 1.0$.

This two-stage efficiency model is most suitable for application in competitive markets, where the firms or DMUs are price takers in both factor and product markets. The output-oriented DEA model corresponding to (2.4) may be analogously specified as follows:

$$\text{Max } R = c' y$$

$$\text{s.t.} \quad \sum_{j=1}^{N} x^*(j)\lambda_j \le x^*(k); \sum_{j=1}^{N} y^*(j)\lambda_j \ge y \qquad (2.5)$$

$$\sum \lambda_j = 1, \lambda_j \ge 0$$

where the competitive input and output price vectors are q and c. Likewise and profit-oriented model can be set up for determining optimal input and output vectors jointly. This two-stage model has two interesting implications. First, the first stage incorporates in its optimal basis the multioutput production frontier specific to each firm j. Thus if B(j) is the optimal nondegenerate basis in (2.1), the optimal output can be written as:

$$y^*(j) = H(j) \, x^*(j); \ H(j) = B(j)^{-1} \qquad (2.6)$$

Here each optimal output is a linear combination of all the optimal inputs as in a productoin frontier. The second stage specification in (2.4) tests if firm k is relatively efficient in relation to all other firms in the industry comprising N firms. Interfirm competition in the competitive market yields the result that only the efficient firms survive. Thus consider the output oriented model (2.5) and let $Y^* = Ny^*$ denote the aggregate optimal output vector. If the market demand vector is D and $Y^*$ exceeds (falls short of) D, then the competitive prices c fall (rise), till the equilibrium $Y^* = D$. Hence the two stages complement each other in securing a Pareto efficient industry with output $y^* = D$ comprising the N efficient firms each producing an output $y^*$. Secondly, the adjustment process in reaching the optimal level may be directly introduced at the individual firm level in the first stage (2.1) and the industry level in the second stage (2.5) for instance. Thus there are two sources of efficiency, one at the firm level and the other at the industry level. Since the industry level adjustment occurs through the competitive market prices denoted by c in (2.5), one could directly introduce a linear industry demand function as

$$c = a - \hat{b}Y; Y = Ny$$

where $\hat{b} = (\hat{b}_r)$ is a diaponal matrix with positive diaponal elements. The objective function in (2.5) is then transformed to a quadratic form

$$R = a'y - Ny'\hat{b} y$$

The optimal solution of the resulting quadratic program then characterizes the adjustment process.

The DEA model for the industry has however more flexibility in a dynamic setting. This may be illustrated with respect to the input oriented model

(2.4) for example. Consider a time-oriented version of the overall efficiency model (2.4), when the inputs $x_t$ are chosen so as to minimize an intertemporal cost function

$$C = \sum_{t=1}^{\infty} \rho^t \left[ (x_t - x_{t-1})' \hat{w} (x_t - x_{t-1}) + (x_t - x_t^d)' \hat{h} (x_t - x_t^d) \right]$$

s.t. $\quad \sum_{j=1}^{N} x_t^*(j)\lambda_{jt} \le x_t; \sum_{j=1}^{N} y_t^*(j)\lambda_{jt} \ge y_t^*(k) \qquad (2.7)$

$$\sum_{j=1}^{N} \lambda_{jt} = 1; \lambda_{jt} \ge 0, j \in I_N$$

Here $\rho$ is a positive rate of discount, $\hat{w}$ and $\hat{h}$ are diagonal matrices with positive diaponal elements and $x_t^d$ denotes the desired input goal for firm k. The objective function of this model (2.7) differs from that of (2.4) in two significant ways. First, it shows the dynamics of the producer's adjustment process over time, e.g., the first component of the quadratic cost indicates the risk aversion of the producer towards input fluctuations over time, whereas the second component reflects the disequilibrium cost due to deviations from the desired goal. We note that this sort of quadratic cost function has been frequently adopted in aggregate production scheduling models, e.g., the HMMS model which has been applied in DEA framework by Kumar and Sinha (1998). Secondly, this intertemporal model can be easily compared with a sequential one period ahead model, when one solves for the optimal solution $x_{t+1}^*$ given the initial solution $x_t$ and then moving t forward. The steady state model may then be compared with the short run model and the implicit cost of myopic decisions may be evaluated.

Denote the optimal solutions of (2.7) by $\lambda_t^* = (\lambda_{jt}^*), x_t^*$. Then it follows by the Pontryagin maximum principle that if $DMU_k$ is efficient then it must satisfy the following necessary conditions which are also sufficient by the strict convexity of the objective function:

$$\rho \hat{w} x_{t+1}^* - (\hat{h} + \hat{w}(1+\rho)) x_t^* + \hat{w} x_{t-1}^* + \hat{h} x_t^d + \beta_t^* = 0$$
$$y_t^*(j)\alpha_t^* = x_t^*(j)\beta_t^* + \beta_{0t}^* \qquad j=1,2,\dots,N \qquad (2.8)$$

where $\alpha_t^*, \beta_t^*$ and $\beta_{0t}^*$ are the optimal values of the Lagrange multipliers. The steady state solution follows from (2.8) as:

$$\overline{x}* = \overline{x}^d + \hat{h}^{-1}\overline{\beta}* \qquad (2.9)$$

The myopic one period ahead optimal solution is:

$$x_t^* = (\hat{w} + \hat{h})^{-1}[\hat{w}\, x_{t-1}^* + \hat{h}\, x_t^d + \beta_t^*] \qquad\qquad (2.10)$$

If the weights are such that $\hat{w} = I$ and $\hat{h} = \hat{v}$ where $\hat{v} = (\hat{v}_i)$ are the variances of input prices or costs, then the optimal conditions (2.9) and (2.10) for an efficient DMU appear as follows:

$$\bar{x}^* = \bar{x}^d + \hat{v}^{-1}\beta^* \text{ (steady state)} \qquad\qquad (2.11)$$
$$x_t^* = \hat{v}^{-1}(x_{t-1}^* + \beta_t^*) + x_t^d \text{ (myopic)}$$

Several interesting differences of the models (2.8) through (2.10) may be noted. First of all, the production frontier equation (2.8) is changing over time as the optimal input vectors $x_t^*$ vary over time. The myopic solution (2.10) ignores the influence of the future optimal inuts $x_{t+1}^*$ and the discount rates. Secondly, the higher the variance of input prices or costs, the lower the optimal levels of the inputs ($x_t^*$ or $\bar{x}^*$) in both the myopic and the steady state. Finally, the convergence of the optimal inputs $x_t^* \to \bar{x}^*$ in the intertemporal model (2.8) to the steady state $\bar{x}^*$ is guaranteed only if the stable root of the second order difference equation (2.8) holds, either through direct managerial policy or by the so-called perfect foresight condition. This aspect has been anlayzed by Sengupta (1999) when some inputs are capital inputs, while others are variable inputs. Finally, note that the discount rate $\rho$ has no effect on the steady state behavior of optimal inputs; but the higher the goal in terms of the target inputs $\bar{x}^d$, the greater the level of optimal inputs $\bar{x}^*$ and $x_t^*$. Thus the differences in relative efficiencies of different DMUs may be attributed in part to their different goals or targets.

## B. Flexibility in Production Process

Two types of adaptive behavior are very important in improving the efficiency of production processes. One is the flexibility reflected in a flatter average cost curve, when the firm faces an uncertain demand which causes price fluctuations. Sometimes the firms have options to choose between two types of technology, fixed and flexible in such modern industries as microelectronics and what is broadly called flexible manufacturing systems (FMS). Both these cases have strategic importance in efficiency measurement, which will be discussed below.

A second type of adaptive behavior for firms arises through learning. Two aspects are important here. One is the cumulative experience gained by workers over time, which is sometimes referred to as 'knowledge capital'. Two forms of this knowledge capital has been frequently used in new growth theory in economics. One is the concept of 'learning by doing' in the capital goods sector due to Arrow (1962). The other is the cumulative number of new designs invented by the whole industry. This is stressed by Lucas (1993), Romer (1990) and others. The sources of efficiency in DEA framework are more clearly brought out through these learning curve effects, particularly in intertemporal frameworks.

Consider first the concept of *flexibility* in production processes measured by a quadratic total cost function

$$TC = \sum_{r=1}^{s} C_r = \sum_r (a_r y_r^2 + b_r y_r) + g$$

Following Marschak and Nelson (1962), DeGroote (1994) has argued that a decrease in the parameter $a_r$ corresponds to an increase in flexibility as reflected in a flatter average cost curve. On using this cost function a DEA model could be set up for optimally determining the output vector $y = (y_r)$ and technical efficiency $\theta$, for every given price vector $p = (p_r)$ of output:

$$Max\pi = \sum_{r=1}^{s} \left[ p_r y_r - (a_r y_r^2 + b_r y_r) - g \right] - \theta$$

$$\text{s.t.} \quad \sum_{j=1}^{N} x_{ij}\lambda_j \leq \theta x_{ik} ; \sum_{j=1}^{N} y_{rj}\lambda_j \geq y_r \tag{2.12}$$

$$\sum \lambda_j = 1, \lambda_j \geq 0 ; \theta \text{ free in sign}$$

Let asterisks denote the optimal solutions with a positive optimal output $y_r^*$ for an efficient firm. Then it follows by Kuhn-Tucker theorem that the optimal outputs and profit are

$$y_r^* = (2a_r)^{-1}(p_r - b_r - \alpha_r^*)$$
$$\pi^* = (4a_r)^{-1}\left[(p_r - b_r)^2 - \alpha_r^{*2}\right] - g \tag{2.13}$$

where $\alpha_r^*$ is the dual variable corresponding to the output constraint. Now assume that the price vector $p$ is stochastic with mean $\bar{p} = (\bar{p}_r)$ and variance $\sigma^2 = (\sigma_r^2)$, where demand uncertainty is assumed to be reflected in market clearing prices. Then on taking expectations of optimal profits in (2.13) one may easily derive

$$E\pi^* = \sum_{r=1}^{s} \left[ (4a_r)^{-1}\left\{ (\bar{p}_r - b_r)^2 + \sigma_r^2 - \alpha_r^{*2} \right\} \right] - g \tag{2.14}$$

This shows that as $\sigma_r^2$ rises expected optimal porfit for the more flexible plan rises relative to that for the less flexible plant.

The role of flexibility in an input-oriented DEA model may be similarly specified in terms of the input oriented cost function $C(x) = \sum_{i=1}^{m} q_i x_i = q'x$, where the input price $q_i$ is assumed to be distributed with mean $\bar{q}_i$ and variance $\sigma_i^2$.

$$\text{Min } C_A = \sum_{i=1}^{m} \bar{q}_i x_i + \phi \sum \sigma_i^2 x_i^2 + g$$

$$\text{s.t. } \sum_{j=1}^{N} x_{ij} \lambda_j \leq x_i; \sum_{j=1}^{N} y_{rj} \lambda_j \geq y_{rk}$$

$$\sum \lambda_j = 1, \lambda_j \geq 0$$

By following the same procedure as in (2.14) one can derive the optimal risk adjusted cost $C_A^*$ as follows

$$C_A^* = (4\phi\sigma_i^2)^{-1}(\beta_i^* - \bar{q}_i)(1 + 2\bar{q}_i) + g \qquad (2.15)$$

Here less $\phi$ implies more flexibility in the cost function and hence as $\sigma_i^2$ increases, the expected optimal risk adjusted cost $C_A^*$ falls relative to that for the less flexible plant, if $\beta_i^* \leq \bar{q}_i$. The latter condition would normally hold by the Kuhn-Tucker necessary condition. Since the parameter $\phi$ may be interpreted as a risk aversion measure, one could identify the presence of unequal degrees of risk aversion among firms as a potential source of inefficiency.

A second concept of flexibility arises through technology in the FMS literature; see e.g., Fine (1986) and Sengupta (1995a). Here flexible capacity is a separate input along with other inputs and this capacity enables a firm to hedge against uncertainty or fluctuations in future demand, but this occurs at the expense of increased costs to investing in flexible manufacturing capacity, as compared with dedicated or nonflexible capacity. Thus we have here a two-stage decision problem for efficiency. In the first stage the firm must make its investment decision in installing manufacturing capacity, before the product demand is known. In the second stage the demand for products are revealed and the firm implements its optimal production decisions.

We assume that each firm can sell s different products indexed by r = 1,2,...,s. It has s+1 types of capacity available: dedicated r-th capacity for r = 1,2,...,s, each of which can manufacture only members of its own product family and flexible capacity (indexed by F) that can produce any or all of the s product families. The cost of any product changeover is infinitely high for dedicated (i.e., nonflexible) capacity, while it is zero for flexible capacity. Denote by $k_r$ and $K_F$ the amounts of dedicated and flexible capacity to be optimally chosen by the firm and let $a_r(K_r)$ and $a_F(K_F)$ be the acquisition costs for these technologies, where $a_F(K) > a_r(K) > 0$ for any positive K.

About uncertain demand we assume there are T possible states of the world, where the state t occurs with probability $w_t$ with $w_t > 0$ and $\Sigma\ w_t = 1$. For each realization of state t the firm chooses production levels for each product family, subject to the capacity constraints of the first stage. Let $y_{tr}$ and $z_{tr}$ be the outputs produced on the dedicated and flexible capacity respectively, when state t occurs and the revenue and cost functions are $p_{tr}(y_{tr} + z_{tr})$ and $c_r(y_{tr}) + c_{rF}(z_{tr})$. Then we can formulate a FMS model with DEA characteristics as follows:

$$\text{Max}\,\pi = -a_F K_F - \sum_{r=1}^{s} a_r K_r + \sum_{t=1}^{T} w_t \sum_{r=1}^{s} \left\{ p_{tr}(y_{tr} + z_{tr}) - c_r(y_{tr}) - c_{rF}(z_{tr}) \right\}$$

$$\text{s.t.} \quad y_{tr} \leq K_r; \quad t = 1,2,\ldots,T; r = 1,2,\ldots,s$$

$$\sum_{r=1}^{s} z_{tr} \leq K_F; \quad t = 1,2,\ldots,T$$

$$\sum_{j=1}^{N} y_{trj}\lambda_{jt} \geq y_{tr}; r = 1,2,\ldots,s$$

$$\sum_{j=1}^{N} z_{trj}\lambda_{jt} \geq z_{tr}; r = 1,2,\ldots,s \qquad (2.16)$$

$$\sum_{j=1}^{N} x_{ij}\lambda_{jt} \leq \theta x_{ik}; i = 1,2,\ldots,m$$

$$\sum_{j=1}^{N} \lambda_{jt} = 1; \lambda_{jt} \geq 0$$

Here the decision variables are the capacity levels ($K_F$, $K_r$; r = 1,2,...,s) and the outputs $y_{tr}$, $z_{tr}$; t = 1,2,...,T and r = 1,2,...,s determined in the first stage, when future demand is only known by its probability $w_t$. In the second stage the optimal outputs of $y_{tr}$ and $z_{tr}$ are compared with their potential maximum outputs

$$\hat{y}_{tr} = \sum_{j=1}^{N} y_{trj}\lambda_{jt}^* \quad \text{and} \quad \hat{z}_{tr} = \sum_{j=1}^{N} z_{trj}\lambda_{jt}^*.$$ Also the input efficiency of firm k (k =

1,2,...,N) is tested by the technical efficiency measure $\theta^*$ at its optimal value.

Clearly if the revenue and cost functions are linear, then this defines a linear programming problem with stochastic features. This model may be presented in more simple terms if the demand uncertainty is specified separately and a chance-constrained relation for the demand constraint $y_r + z_r \geq d_r$ is adjoined separately as follows:

$$\text{Max}\,\pi = -a_F K_F - \sum_r a_r K_r + \sum_{r=1}^{s} \left\{ p_r (y_r + z_r) - c_r (y_r) - c_{rF}(z_r) \right\} - \theta$$

$$\text{s.t. } y_r \leq K_r; \sum_{r=1}^{s} z_r \leq K_F; \; r = 1,2,\ldots,s$$

$$\sum_{j=1}^{N} y_{rj}\lambda_j \geq y_r; \sum_{j=1}^{N} z_{rj}\lambda_j \geq z_r; r = 1,2,\ldots,s$$

$$\sum_{j=1}^{N} x_{ij}\lambda_j \leq \theta x_{ik}; \; i = 1,2,\ldots,m \qquad (2.17)$$

$$\sum_{j=1}^{N} \lambda_j = 1; \lambda_j \geq 0$$

and $y_r + z_r = \overline{d}_r - q_r \sigma_r; q_r = F^{-1}(u_r); \; r = 1,2,\ldots,s$

Here $u_r$ is the probability with which the demand constraint is satisfied and each demand $d_r$ is assumed to be distributed with mean $\overline{d}_r$ and variance $\sigma_r^2$.

Several features of this stochastic model (2.16) or (2.17) may be noted. First of all, the more the fluctuations or uncertainty in demand, the more profitable it is to invest in flexible capacity. Thus several domains in the capacity region may be determined in the optimal phase, e.g., only fixed or dedicated capacity is used, both fixed and flexible capacities are used or only the flexible capacity is used. Secondly, the reference firm or $DMU_k$ satisfies two-stage efficiency only if it uses the optimal capacities, optimal outputs and the highest level of input efficiency, i.e., $\theta = 1.0$. Thus the static DEA models which ignore the decision problem of choosing optimal capacities under demand uncertainty would indicate a biased measure of efficiency. Finally, the fluctuations in demand measured by the variance $\sigma_r^2$ would affect the optimal choice of both capacity and output variables. In product markets where the technology competition is very intensive in modern times, this stochastic aspect of demand changes may be a very important determinant of optimal input and output efficiency.

## C. Learning and Cumulative Experience

Arrow's learning by doing models the technical innovation process of the whole economy on the example of labor productivity influenced by the cumulative experience of the airframe industry. Here experience is measured by the cumulated gross investment $(I(v))$

$$K(t) = \sum_{-\infty}^{t} I(v)dv \qquad (2.18)$$

and the aggregate production function is specified as

$$(t) = F[K(t), A(t) L(t)] \qquad (2.18.1)$$

where the current efficiency of labor (i.e., labor productivity) is measured by

$$A(t) = K(t)^{\gamma}, 0 < \gamma < 1 \qquad (2.18.2)$$

It follows that even if the production function $F(K, AL)$ has constant returns to scale in the two inputs $K$ and $AL$ as in the neoclassical model, the overall production function exhibits increasing returns to scale.

On using the specifications (2.18.1) and (2.18.2) for a single output and single capital input $(K_j)$ one could set up an overall efficiency model in a DEA framework as follows:

$$\text{Min}C = cK + \sum_{i=1}^{m} \hat{c}_i x_i$$

$$\text{s.t.} \quad \sum_{j=1}^{N} K_j^{\gamma_j} x_{ij} \lambda_j \le K^{\gamma} x_i; \ I = 1,2,\ldots,m \qquad (2.18.3)$$

$$\sum_{j=1}^{N} K_j \lambda_j \le K; \sum_{j=1}^{N} y_j \lambda_j \ge y_k$$

$$\sum_j \lambda_j = 1; \lambda_j \ge 0$$

Here the decision variables are the $m+1$ inputs $x_1, x_2, \ldots, x_m$ and capital $K$ and the $n$ weights $\lambda_j$. Clearly this defines a nonlinear programming problem. By the duality theorem in nonlinear programming it follows that if firm $j$ is overall efficient, it must satisfy the necessary condition

$$\alpha^* y_j = \sum_{i=1}^{m} \beta_i^* A_j x_{ij} + b^* K_j + \beta_0^* \qquad (2.18.4)$$

where $A_j = K_j^{\gamma_j}$ and $\alpha^*, \beta_i^*, b^*, \beta_0^*$ are the optimal Lagrange multipliers for the constraints. If $\gamma_j = 0$, this reduces to the standard production frontier in the DEA model. If $K_j$ enters in a simple multiplicative form to the inputs and the firm j is efficient, then the production frontier takes a simpler form as

$$\alpha^* y_j = \sum_i \beta_i^* \gamma_j K_j x_{ij} + b^* K_j + \beta_0^*$$

The effect of $K_j$ is to augment the productivity parameter $\beta_i^*$ of each input $x_{ij}$. Note however that we need data on $K_j$ and its parameter $\gamma_j$ either on a priori basis or from past experience of learning curve effects.

As a second case consider human capital in the form of knowledge gained through experience. Let $s_j$ be the firm specific knowledge (capital) in firm j and $S = \Sigma\, s_j$ represents knowlede capital for the whole industry. Then we assume like Lucas (1993) that the learning process of knowledge evolution follows the difference equation over time as

$$\Delta s_j(t) = \delta_j S^\gamma(t) a_j s_j(t); \; j = 1, 2, \ldots, N \tag{2.19.1}$$

Here $a_j$ is the fraction of workers' time spent on learning, $\gamma$ is the industry effect and $\delta_j$ is the joint effect of $s_j(t)$ and $S(t)$. Given time series data of the past experiences of firms, the equation (2.19.1) can be estimated and then one can use the estimated values $\Delta\hat{s}_j$ to solve for optimal levels of inputs $x_i$ and knowledge capital as by the following DEA model which minimizes total costs:

$$\text{MinC} = \sum_{i=1}^m c_i x_i + c_0 \Delta s$$

$$\text{s.t.} \quad \sum_{j=1}^N x_{ij}\lambda_j \le x_i; \sum_{j=1}^N \Delta s_j \lambda_j \le \Delta s \tag{2.19.2}$$

$$\sum_{j=1}^N y_{rj}\lambda_j \ge y_k; \sum \lambda_k = 1; \lambda_j \ge 0$$

For the efficient firm k we obtain the production frontier as follows:

$$\sum_{r=1}^s \alpha_r^* y_{rk} = \sum_{i=1}^m \beta_i^* x_{ik} + \beta_0^* + \hat{\delta}_k S^{\hat{\gamma}}(t) \hat{a}_k s_k(t)$$

where a hat denotes the estimated values of parameters from equation (2.19.1).

Two flexible features of this formulation have to be noted. First, the industry effect on firm-specific productive efficiency can be directly captured here; hence the impact of knowledge spillover and diffusion of technical knowhow may be directly introduced as sources of efficiency. Secondly, the regression approach in (2.19.1) can be directly applied to estimate the nonlinear feature of learning through diffusion, while the standard DEA model can be applied as in (2.19.2) to estimate the production frontier. This two-stage regression-cum DEA approach can also be applied in the earlier case (2.18.3).

As a final example of learning we consider the learning curve effect which postulates that costs decline as cumulative output increases. Norsworthy and Jang (1992) in their analysis of technological change and its impact on productivity have found this type of learning by doing to be a major contributor of improvement of manufacturing efficiency in the modern process and product technology-intensive industries such as the semiconductor, microelectric and telecommunication industries. To capture this learning curve effect we write the cost function in a separable form as $C = \Sigma C_r$, where

$$C_r = y_r^{-\phi_r}, \phi_r > 0 ; r = 1,2,\ldots,s$$

and set up the DEA model as in (2.12) but in a cost minimizing form

$$\text{Min} C_T = \sum C_r + \theta$$

$$\text{s.t.} \quad \sum_{j=1}^{N} x_{ij} \leq \theta x_{ik}; \sum_{j=1}^{N} y_{rj}\lambda_j \geq y_r \qquad (2.20.1)$$

$$\sum \lambda_j = 1; \lambda_j \geq 0 ; \theta \text{ free in sign}$$

By Kuhn-Tucker theorem the optimal outputs $(y_r^*)$ and the weights $\lambda_j^*$ must satisfy the following conditions:

$$\phi_r / y_r^{*1+\phi_r} \leq \alpha_r^*$$

$$\sum_{i=1}^{m} \beta_i^* x_{ik} = 1; \sum_{r=1}^{s} \alpha_r^* y_{rj} \leq \sum_i \beta_i^* x_{ij} - \beta_0^* \qquad (2.20.2)$$

If the k-th firm or $DMU_k$ is efficient and equality holds in the first condition of (2.20.2), we obtain for the optimal average cost

$$AC_r^* = C_r^* / y_r^* = \alpha_r^* / \phi_r$$

which shows that average cost declines as the learning parameter $\phi_r$ increases. This leads to higher optimal output. We note that this learning curve effect is most closely associated with R&D inputs which comprise a set of indivisible inputs much different from the conventional physical inputs.

We may conclude this section by noting that the impact of market competition on firm efficiency is not directly analyzed in traditional DEA models. That is why Athanassopoulos and Thanassoulis (1995) in their DEA application to private sector market-oriented enterprises suggested a two-stage process of evaluation of overall efficiency. The first stage relates to the firm attracting demand and generating revenue. This stage is called *'market efficiency'*. Advertising expenditure and marketing inputs are utilized in this stage to optimally influence the market demand. The second stage minimizes total operating costs for a given revenue level and it is termed *'cost efficiency'*. Two major implications of this two-stage approach are as follows. One is the demand fluctuations that influence the output prices. For DEA model (2.12) for example the prices $p_r$ would not be constant but would vary according to total market demand. Secondly, cost efficiency is viewed conditional on total revenue determined by market demand in the first stage, where market demand is partly influenced by alternative market strategies in respect of various types of advertising expenses.

## 2.2    Market and Cost Efficiency Under Competition

Competitive pressure in market demand tends to improve the cost efficiency of individual firms in an industry. One way to model this efficiency improvement in a single output case is to assume that the competitive industry faces a demand (D) function as

$$D = a - bp$$

Each firm faces the identical cost function

$$C_j = c_0 + c_1 y_j + c_2 y_j^2; j = 1,2,\ldots,N$$

We assume free entry and an unlimited number of potential entrants. So long as average cost ($AC_j$) is lower than marginal cost ($MC_j$) the rule of $p = MC$ would generate excess profits, which would invite new entry till the condition of zero profits is reached for each firm at $p = AC_{min}$. Thus we minimize average cost

$$AC = c_0/y + c_1 + c_2 y; c_0,c_1,c_2 > 0$$

at the output level

$$y^* = \sqrt{c_0 / c_2}$$

and obtain

$$p = AC_{min} = 2\sqrt{c_0 / c_2} + c_1$$

This yields the equilibrium output Y* and price p* for the competitive industry as

$$Y^* = D = a - b\left(c_1 + 2\sqrt{c_0 / c_2}\right)$$
$$p^* = c_1 + 2\sqrt{c_0 / c_2}, N = Y^* / y^*$$

For instance let $c_0 = 200$, $c_1 = 10$, $c_2 = 2$ and a $= 800$, b $= 8.0$. then

$$y^* = y_j^* = 10; p^* = 50; Y^* = D = 400; N = 40$$

We can easily adopt this competitive industry model of equilibrium output in terms of the allocative efficiency formulation of the DEA model, where each firm j produces one output ($y_j$) by using m inputs ($x_{ij}$). Thus in order to test the overall efficiency of k-th firm one sets up the following quadratic programming (QP) model:

$$\text{Min } AC + \theta = (c_0 / y) + c_1 + c_2 y$$

$$\text{s.t.} \quad \sum_{j=1}^{N} X_j \lambda_j \leq \theta X_k; \sum_{j=1}^{N} y_j \lambda_j \geq y; \sum \lambda_j = 1$$

and
$$\lambda_j \geq 0; j = 1, 2, \ldots, N$$

where $X_j$ is the m-element column vector of positive inputs for firm j and the cost coefficients $c_0, c_1, c_2$ are assumed to be known or estimated. Here one can apply the regression method to estimate these cost parameters. Based on the Lagrangean function

$$L = -\left(c_0 / y + c_1 + c_2 y\right) - \theta + \beta'\left(\theta X_k - \sum_j X_j \lambda_j\right)$$
$$+ \alpha\left(\sum_j \lambda_j y_j - y\right) + \beta_0 \left(1 - \sum \lambda_k\right)$$

One could derive the necessary condition for firm k to be efficient as

$$y_k = y^* = \left(\beta_0^* / \alpha^*\right) + \left(\beta^{*\prime} / \alpha^*\right) X_k$$

$$\beta^{*'}X_k = 1; AC_{min} = (c_0/y^*) + c_1 + c_2y^*$$

With the market demand function as

$$Y^* = Ny^* = a - bp \text{ and } p = AC_{min}$$

one obtains the equilibrium industry output as

$$Y^* = a - b \left[c_0/y^* + c_1 + c_2y^*\right]$$

Clearly other forms of market competition e.g., imperfect competition can be analyzed in an analogous fashion. For example consider the allocative efficiency model where the optimal output $y^*$ is chosen for each firm by maximizing its prfots $\pi$, when the market demand is given as

$$P = a - b \, Ny$$

i.e.

$$\text{Max } \pi = py - c_0 - c_1y - c_2y^2$$
$$\text{s.t.} \quad \sum_{j=1}^{N} X_j\lambda_j \le X_k; \sum_{j=1}^{N} y_j\lambda_j \ge y$$
$$\Sigma \lambda_j = 1, \lambda_j \ge 0; j = 1,2,\ldots,N$$

In this case the efficiency conditions are

$$2(bN + c_2)y^* > a - c_1 - \alpha^*$$
$$\alpha^* y_j \le \beta_0^* + \beta^{*'} X_j$$

If firm k is efficient in the sense of maximum profits, then $y_k = y^*$ and

$$\alpha^* y_k = \beta_0^* + \beta^{*'} X_k; y^* = (a - c_1 - \alpha^*)[2(c_2 + bN)]^{-1}$$

The influence of N on the optimal output level $y^*$ can be directly estimated here and the source of output inefficiency due to the divergence of $y_j$ from $y^*$ can be identified.

We consider now allocative efficiency under conditions of imperfect competition. We set up first a DEA model for testing the technical efficiency of a reference unit k in a cluster of N units, where each unit j or DMU$_j$ has m inputs $(x_{ij})$ and s outputs $(y_{rj})$:

Min θ

s.t.     $\sum_{j=1}^{N} x_{ij}\lambda_j \leq \theta x_{ik}$; $I = 1,2,\ldots,m$

$\sum_{j=1}^{N} y_{rj}\lambda_j \geq y_{rk}$; $r = 1,2,\ldots,s$                      (2.2.1)

$\sum_{j=1}^{N} \lambda_j = 1; \lambda_j \geq 0; \theta \geq 0$

In vector matrix form this is:

Min θ, s.t. $X\lambda \leq \theta X_k$; $Y\lambda \geq Y_k$; $\lambda' e = 1$; $\lambda \geq 0$                      (2.2.2)

Where e is a column vector with N elements, each of which is unity and the prime denotes a transpose. Here the input $(X_j)$ and output $(Y_j)$ vectors $(j=1,2,\ldots,N)$ are all observed and this is called an input oriented model in DEA literature. Here the reference unit k is compared with the other (N-1) units in the cluster. Let $\lambda^* = (\lambda_j^*)$ and $\theta^*$ be the optimal solutions of the above DEA model with all the slack variables zero. Then the reference unit k or $DMU_k$ is technically efficient if $\theta^* = 1$ and the first two sets of inequalities in (2.2.1) hold with equality. Thus the optimal value of $\theta^*$ provides a measure of technical efficiency (TE). If $\theta^*$ is positive but less than unity, then it is not technically efficient at the 100 percent level. Overall efficiency $(OE_j)$ of a DMU or firm j however combines both technical $(TE_j)$ or production efficiency and the allocative $(AE_j)$ or price efficiency as follows:

$OE_j = TE_j \times AE_j$ ; $j = 1,2,\ldots,N$                      (2.2.3)

To characterize overall efficiency of the reference unit $DMU_k$ one sets up the linear programming (LP) model as follows:

$\underset{x,\lambda}{\text{Min}} q'x$   s.t. $\sum_{j=1}^{N} X_j\lambda_j \leq x$

$\sum_{j=1}^{N} Y_j\lambda_j \geq Y_k$                      (2.2.4)

$\lambda'e = 1; \lambda \geq 0; x \geq 0$

Here q is an m-element vector of unit costs or input prices as observed in the competitive market and x is an input vector to be optimally decided by $DMU_k$ along with the weights $\lambda_j$. Here $X_k$ and $Y_k$ are the observed input and output vectors for the reference unit k, whereas x is the unknown decision vector to be optimally determined. Let $\lambda^*$ and $x^*$ be the optimal solution of the LP model (2.2.3) with all slacks zero. Then the minimal input cost is given by $c_k^* = q'x^*$, whereas the observed cost of the reference unit is $c_k = q'X_k$. Hence the three efficiency measures are defined as follows:

$$TE_k = \theta^*; OE_k = c_k^* \text{ and } AE_k = OE_k / TE_k \qquad (2.2.5)$$

Note that the input vector x in (2.2.4) is a decision vector to be optimally chosen, whereas $X_k$ is the observed data in (2.2.1). If $\theta^*X_k = x^*$, then the two models generate identical optimal solutions; otherwise the two optimal solutions are very different. The dual problems corresponding to (2.2.4) and (2.2.1) appear as follows:

$$\text{Max } \alpha' Y_k + \alpha_0$$
$$\text{s.t.} \quad \beta \leq q \text{ and } \beta' X_j \geq \alpha Y_j + \alpha_0; j=1,2,\ldots,N \qquad (2.2.6)$$
$$\alpha, \beta \geq 0, \alpha_0 \text{ free in sign.}$$

Let asterisks denote optimal values and let $DMU_k$ be efficient. Then it must follow from (2.2.6) that the production frontier for the k-th unit is as follows:

$$\alpha^{*'} Y_k = \beta^{*'} X_k - \alpha_0^*$$

but since $\beta^*$ is constrained as $\beta^* \leq q$, we must have $\alpha^{*'} Y_k \leq q' X_k - \alpha_0$. Thus so long as the actual inputs $X_k$ are not equal to their optimal levels $x^*$, this efficiency gap measured by $(q' X_k - \alpha_0^* - \alpha^{*'} Y_k)$ may persist. Thus the constraint $\beta^* < q$ reflects the fact that the observed input $X_k$ of the reference unit may or may not be equal to the optimal level $x^*$, when all firms face the same competitive price q. There is no such constraint for the dual problem (2.2.7). Note that if $\alpha_0^*$ is positive (negative or zero), then we have increasing (decreasing or constant) returns to scale.

We now consider a more generalized version of the overall efficiency model (2.2.4) where both input (q) and output prices (p) are assumed to be available and the optimal vectors of input and output are optimally chosen as follows:

$$\text{Max } p' y - q' x$$
$$\text{s.t.} \quad \sum_{j=1}^{N} X_j \lambda_j \leq x; \sum_{j=1}^{N} Y_k \lambda_j \geq y$$
$$x \leq X_k; y \geq Y_k; \lambda' e = 1 \qquad (2.2.8)$$

$$x, y, \lambda \geq 0$$

Here $(x,y)$ are the control vectors of inputs and output to be optimally determined and $(X_k, Y_k)$ denote the observed levels for $DMU_k$. The dual of this problem then becomes

$$\text{Max } v' Y_k - u' X_k - \alpha_0$$
$$\text{s.t.} \quad p \leq \alpha - v, \ q \geq \beta - u \qquad\qquad (2.2.9)$$
$$\beta' X_j \geq \alpha' Y_j + \alpha_0, \ j = 1, 2, \ldots, N$$
$$(u, v, \alpha, \beta) \geq 0, \ \alpha_0 \text{ free in sign}$$

Two special cases of the genealized model (2.2.8) are of great importance. One is the simpler output-oriented model where demand $(d_r)$ for output $(y_r)$ is subject to a probability distribution $F(d_r)$ and the objective function is to maximize the expected value of total revenue minus expected inventory cost. This yields the model

$$\text{Max } E\left[ \sum_{r=1}^{s} p_r \min(y_r, d_r) - \sum_{r=1}^{s} h_r (y_r - d_r) \right] \qquad (2.2.10)$$
$$\text{s.t. } X\lambda \leq X_k; \ X\lambda \geq y, \ \lambda' e = 1, \lambda \geq 0$$

where $\lambda$ and $y$ are the unknown vectors to be optimally solved for and $h_r$ is the observed unit cost of positive inventory for $y_r > d_r$. Denoting optimal values by asterisks, the efficient $DMU_k$ would then satisfy the following marginal condition:

$$F(y_r^*) = (p_r + h_r)^{-1}(p_r - \alpha_r^*)$$
$$\alpha^{*'} Y_k = \beta^{*'} X_k - \alpha_0^*$$

Clearly higher output pice and lower inventory costs would increase the optimal output levels $y_r^*$ which may be compared with the observed outputs $y_{rk}$ in output vector $Y_k$.

The second case is an input-oriented model, where the input decisions $x_i$ are equal to planned values $\bar{x}_i$ plus an error term $\varepsilon_i$ with a zero mean and fixed variance. The errors are disturbances such as mistakes or unexpected difficulties in implementing a planned value $\bar{x}_i$. The planned values $\bar{x}_i$ are the decision variables which have to be optimally chosen by each DMU and the error process $\varepsilon_l$ is realized after the planned value of $x_i$ is optimally selected. The input constraints now turn out to be chance constrained

$$\text{Prob}\left[\sum_{j=1}^{N} x_{ij}\lambda_{ij} \le \overline{x}_i + \varepsilon_i\right] = \delta_i, 0 < \delta_i < 1$$

where $\delta_i$ is the tolerance level of the i-th input constraint. The simpler model then takes the following form:

$$\text{Min} \sum_{i=1}^{m} q_i \overline{x}_i$$

$$\text{s.t.} \quad \sum_{j=1}^{N} x_{ij}\lambda_j = \overline{x}_i + w_i; w_i = F^{-1}(1 - \delta_i) \quad\quad (2.2.11)$$

$$\sum_{j=1}^{N} y_{rj}\lambda_j \ge y_{rk}; \lambda'e = 1, \lambda \ge 0$$

$$i=1,2,\ldots,m; r=1,2,\ldots,s$$

Clearly the input uncertainty is here captured by the term $w_i$ which depends on the level $\delta_i$ of chance constraint, e.g., the higher the level of $w_i$, the lower would be the optimal palnned inputs $\overline{x}_i^*$.

We consider now the role of market competition in the efficiency framework. Hence we assume that each firm or $\text{DMU}_j$ produces a single homogeneous output denoted by $y_j$, where the total industry output is denoted by $y_T = \sum_{j=1}^{N} y_j$. If N is large and the firms or DMUs are competitive, then the output price p is a constant, unaffected by the size of each individual firm. In this case the price can be viewed as $p = \overline{p} + \varepsilon$ made up of two components: the expected price $\overline{p}$ and a random part $\varepsilon$ with a zero mean and a constant variance $\sigma_\varepsilon^2$. The total cost of inputs for each firm may now be related to output as

$$c(y_j) = c\, y_j + F_j$$

Assuming a linear form, where $F_j$ is the fixed cost and c is marginal cost that is assumed to be identical for each firm. Maximization of expected profits would then yield the LP model:

$$\text{Max } \overline{\pi} = (\overline{p} - c)y - F$$

$$\text{s.t.} \quad \sum_{j=1}^{N} X_j\lambda_j \le X_k; \sum_{j=1}^{N} y_j\lambda_k \ge y; \lambda'e = 1, \lambda \ge 0 \quad\quad (2.2.12)$$

where y is the unknown decision variable to be optimally selected. In case the market is imperfectly competitive, the price variable then depends on the output supply of different firms. In the homogeneous output case the firms are all alike, and the inverted demand function can be written as:

$$\bar{p} = a - bY_T, \quad Y_T = \sum_{j=1}^{N} y_j; \quad y_k = y$$

The LP model (2.2.12) would then yield the following optimality conditions:

$$(a - c) - b\,Y_T - b\,y* - \alpha* \leq 0$$
$$\alpha*y_j - \beta*'\,X_j - \alpha_0^* \leq 0$$
$$\alpha*, \beta* \geq 0, \quad \alpha_0^* \text{ free in sign}$$

If firm k is efficient, then one must have

$$y* = (a_1 / b) - Y_T - \frac{\alpha*}{b}; \quad y* > 0$$
$$a_1 = a - c > 0$$

where $y* = y_k^*$ is the efficient output of the k-th firm. If all firms are efficient,

then $Y_T^* = \sum_{j=1}^{N} y_j^*$ and one obtains

$$Y_T^* = (N/b)(1+N)^{-1}[a_1 - \alpha*] \qquad (2.2.14)$$

Now we introduce organization slack denoted by $s_k$ in the cost function

$$c(y) = (c + s_k)\,y + F_k; \quad s_k \geq 0$$

Recently Selten (1986) has interpreted this slack concept due to Leibenstein's (1966) X-efficiency as a part of the cost function and introduced a 'strong-slack' hypothesis which maintains that this type of slack has a tendency to increase so long as profits are positive, i.e., this slack can be reduced only under the threat of losses. Including this slack-ridden cost into the profit maximization model would yield the optimality condition for the efficient output as:

$$y* = (a_1 - \alpha - s_k)/b - Y_T; \quad y* > 0$$

where $y* = y_k^*$ is the efficient output of the firm k. If all firms are efficient then

$$Y_T^* = (N/b)(1+N)^{-1}[a_1 - \bar{s} - \alpha^*]$$
$$\bar{p} = a + N(N+1)^{-1}(\alpha^* + \bar{s} - a_1) \qquad (2.2.15)$$
$$\pi_k^* = (\bar{p} - c\,s_k)y^* - F_k$$

when $\bar{s} = \sum_{k=1}^{N} s_k/N$ is the average rate of slack. Several implications follow from this set (2.2.15) of efficiency conditions. First of all, the long run pressure of competition would tend to lead to zero profits $\pi_k^* = 0$ for all k=1,2,...,N according to the strong slack hypothesis. In this case the expected price becomes $\bar{p} = c + \bar{s} + (F/y^*)$. This shows that fixed costs have a strong positive role in determining the long run equilibrium price. The higher the average slack rate $\bar{s}$, the higher is the equilibrium expected price. Secondly, as the number N of firms increases, it increases the volume of total industry output $Y_T^*$ and reduces the average price. Finally, as the average slack rate $\bar{s}$ rises (falls), it increases (decreases) the equilibrium price. Note that in case of weak slack hypothesis all profits are not squeezed out and there remains a divergence of individual ($s_k$) from the average slack rate ($\bar{s}$), when the latter is positive. Thus some inefficency may persist due to the existence of a positive slack.

So far we have assumed that the expected price $\bar{p}$ is the market clearing price equating market demand and supply. If however this is not the case, then the supply y would differ from demand d, where demand is subject to random fluctuations around the mean level $\bar{d}$. In this framework we have to add to the cost function the costs of inventory and shortage C(y – d). Assuming this cost to be quadratic one may then formalize the decision model

$$\text{Max } \bar{\pi} = (\bar{p} - c - s_k)y - (1/2)\gamma E(y-d)^2 - F \qquad (2.2.16)$$
s.t. the same constraints as in (2.2.12).

In this case the optimality conditions for the efficient output becomes

$$y^* = (b+\gamma)^{-1}[(a_1 - \alpha^* - s_k) + \gamma\bar{d} - bY_T] \qquad (2.2.17)$$

where $\gamma$ is the unit cost of inventory or shortage and $\bar{d}$ is the expected level of demand. In this case the marginal impact $\partial y^*/\partial \gamma$ of inventory/excess costs may be either negative or positive according as

$$b(Y_T + \bar{d}) > \text{or}, < (a_1 - \alpha^* - s_k)$$

Again this explains the persistence of some inefficiency, when demand is uncertain and the firm chooses its optimal output by the quadratic criterion of adjusted profits. Furthermore, the higher the mean demand the greater the optimal level of efficient output y*. In case of perfect competition with each firm a price taker, the optimality condition (2.2.17) reduces to

$$y* = \overline{d} + (1/\gamma)(\overline{p} - c - s_k - \alpha*) \qquad (2.2.18)$$

which shows unequivocally that higher inventory costs ($\gamma$) lead to lower optimal output. Again by comparing the observed output $y_k$ with the optimal output y*, one could evaluate the impact of inefficiency. Note that we still have the comparative static results: $\partial y*/\partial \overline{p} > 0$ and $\partial y*/\partial s_k < 0$. Since $\overline{p} = \gamma(y* - \overline{d}) + c + s_k + \alpha*$, we have the results:

$$\overline{p} > MC_T, \text{if } y* > \overline{d}$$

and

$$\overline{p} < MC_T, \text{if } y* < \overline{d} \qquad (2.2.19)$$

where $MC_T = c + s_k + \alpha*$ is total marginal cost with three components: production costs (c), cost of slack ($s_k$) and the cost of discrepancy of observed from optimal output ($\alpha*$). Clearly the case of multiple outputs can be handled in a symmetrical way.


## 2.3  A Two-stage Model of Cost Efficiency

The two-stage efficiency model solves for an optimal revenue net of advertising costs in the first stage, when the market demand characteristics are directly incorporated into the DEA model. Given this optimal net revenue the second stage minimizes the total operating costs so as to determine the cost efficient unit or firm. Clearly under conditions of market uncertainty and imperfect competition, a firm may not be fully market efficient in the first stage, although it may turn out to be cost efficient in terms of the DEA model in the second stage. The second stage may also be given a dynamic interpretation over time in terms of the adjustment cost theory. For example the firm may add an adjustment cost to the overall objective of minimizing operating costs, where the adjustment cost may reflect the costs of deviation sfrom the optimal revenue and its implied input requirements determined in the first stage. Recently Sengupta (1995a) have used this adjustment cost approach in a quadratic from to capture the producer's risk averse attitude towards input and output fluctuations.

Consider first the market efficiency model in a DEA framework, where we assume for simplicity one output but m advertisement inputs ($a_i$) with unit costs $c_i$. The firm maximizes net revenue:

$$\text{Max NR} = pf_k My - \sum_{i=1}^{M} c_i A_i$$

$$\text{s.t.} \quad \sum_{j=1}^{N} A_{ij}\mu_j \leq A_i; \sum_{j=1}^{N} y_j\mu_j \geq y \tag{2.3.1}$$

$$\Sigma \mu_j = 1, \mu_j > 0; j=1,2,...,N$$

by choosing the optimal levels of output (demand) y and the advertisement inputs of different types. Here p is the output price which depends on both demand y facing the firm and its advertisement costs, i.e., $R = py = R(y, A_1,...,A_2)$ where R is gross revenue. This type of model is most suitable for stores under a grocery chain, restaurants or branches of different banks, where $f_k$ is the fraction of customers out of the potential total M of customers who buy from the store k, whose optimal levels of advertisement $A_i$ and demand y are to be determined. Let $A^* = (A_i^*)$ and $y^*$ be the optimal solutions of the LP model (2.3.1) with a given M. Then the unit or firm k has market efficiency at 100% level if the following conditions hold:

$$\frac{Mf_k R^*}{A_i^*}\left(\varepsilon_{R\cdot A_i} + f_f \cdot A_i\right) + \beta_i^* - c_i = 0$$

for i=1,2,...,m

$$p^* f_k M\left(1 - \frac{1}{\varepsilon_{y\cdot p}}\right) = \alpha^* \tag{2.3.2}$$

$$A_k'\beta^* = \alpha^* y - \alpha_0^*; (\alpha,\beta) \geq 0; \alpha_0 \text{ free in sign}$$

where the Lagrangean expression is

$$L = p(y, A)f_k My - c'A + \beta'(A - \sum_j A_j\mu_j) + \alpha(\Sigma y_j\mu_j - y) + \alpha_0(1 - \Sigma\mu_j)$$

and the term $\varepsilon_{z\cdot x}$ denotes the elasticity of z with respect to x, e.g., $\varepsilon_{f_k\cdot A_i}$ denotes the elasticity of the market share $f_k$ with respect to the advertisement expenditure $A_i$, i.e., $\varepsilon_{f_k\cdot A_i} = \frac{\partial f_k}{\partial A_i} \cdot \frac{A_i}{f_k}$. Note that the market efficiency concept here involves three components, e.g., (1) the optimal levels of advertisement expenditures (or marketing inuts), which incorporate the response of market share and market revenue, (2) the optimal levels of demand or output y* of the k-th firm, so that if the observed output $y_k$ is not equal to y* then we have market inefficiency and finally (3) an optimal marketing response frontier, where the optimal firm demand depends on the shadow prices of the marketing input-mix. Thus the market inefficiency may be decomposed into these three levels. Note that in perfect competition one can drop out all the elasticity terms and then obtain for the efficient unit k:

$$\beta_i^* = c_i; p^* f_k M = \alpha^*; A_k' \beta^* = \alpha^* y^* - \alpha_0^*$$

Clearly this is the condition that price equals marginal cost.

Note that the first stage decision model allows the firm to expand the market to an optimal size. This is a very important consideration in the world market today, where the openness in trade and liberalization have expanded the world export market and some firms have achieved phenomenal success in achieving strong economies of scale, which can keep production costs down and also increase the productivity of inputs.

This implies that the firms which achieve market efficiency may profitably exploit methods of reducing total operating costs (TOC) in the short run and the capacity costs in the long run. For the short run cost minimization problem one may extend the DEA model (2.1.4) by adjoining a vector of research inputs ($R_j$) for each firm j

$$\text{Min TOC} = q' x + C(R)$$
$$\text{s.t.} \quad \sum_{j=1}^{N} X_j \lambda_j \leq x; \sum_j R_j \lambda_j \leq R \tag{2.3.4}$$
$$\sum_j Y_j \lambda_j \leq Y_k; \lambda' e = 1, \lambda \geq 0$$

where it is assumed that maximum revenue for firm k is predetermined in the first stage. Here C(R) denotes the total cost of research inputs. If the research cost is linear $C(R) = \Sigma g_i R_i$, then the LP model (2.3.4) may be used for determining the efficient levels of inputs $x_i^*$ and $R_w^*$ and the efficient level of the output vector y*. Then firm k's cost inefficiency may be measured in terms of the gaps: $X_k \neq x^*$, $R_k \neq R^*$ and $y^* \neq d$. But the research inputs tends to lower the input costs of production. To capture this aspect one may assume an input cost function, where the R&D inputs tend to lower the initial unit production cost $q_i$, i.e.,

$$\text{TOC} = \sum_{i=1}^{m} (q_i - f_i) x_i + \sum_{w=1}^{W} h_w R_w$$

Here $f_i$ is the marginal reduction of unit production cost $q_i$ and $h_w$ is the observed cost of the research input $R_w$. This type of cost-reducing process innovation through R&D has been frequently used in a quadratic form in a Cournot-type cooperative R&D model. In case of the long run the output vector $y_t$ has to be constrained at each time point by the capacity vector $\bar{y}$ where the cost of capacity $C(\bar{y})$ has to be adjoined to the objective function. This yields the dynamic TOC minimization model

$$\text{Min TOC} = \sum_{t=1}^{T} \left[ q_t' x_x + C(R_t) + C(\bar{y}_t) + C(y_t) + C(I_t) \right]$$

$$\text{s.t.} \quad \sum_{j=1}^{N} X_{jt}\lambda_{jt} \le x_t ; \sum_j R_{jt}\lambda_{jt} \le R_t$$

$$\sum_j Y_{jt}\lambda_{jt} \le y_t ; y_t \le \bar{y}_t$$

$$I_t = I_{t-1} + y_t - d_t ; \quad t=1,2,\ldots,T$$

$$\lambda_t' e = 1, \lambda_t \ge 0, I_0 = I_T = 0; I_t \ge 0$$

Here $C(y_t)$ and $C(I_t)$ are the production and inventory costs and it is assumed that demand ($d_t$) cannot be backlogged. Clearly we need the condition

$$\sum_{t=1}^{\tau} \bar{y}_t \ge \sum_{t=1}^{\tau} d_t, 1 \le \tau \le T$$

for a feasible and hence an optimal solution to exist. This generalized model (2.3.5) is closely related to the optimal production scheduling model known as the HMMS model in operations research literature, where the cost functions $C(y_t)$, $C(\bar{y}_t)$, $C(I_t)$ are usually assumed to be quadratic in form. The only difference is that it is modified so as to include the DEA framework for comparative performance. Taking the above cost functions in a linear form, one could determine from the above LP model a time profile of operating cost frontier having several components of efficiency, e.g., cost efficiency due to inputs, outputs, capacity and inventory.

## 2.4.    Cost Uncertainty and Capacity Utilization

Capacity utilization has two basic roles in industrial price and output policies. the first is one of the basic propositions in macroeconomics which says that price inflation accelerates as capacity and resource utilization moves higher. The second is the intertemporal implication of changes in capacity inputs, which affect both the fixed and variable costs in the short run. Since every short run productoin and cost function is conditional on a fixed supply of capacity inputs, the short run cost minimization model may not ordinarily yield the long run cost frontier. We consider here first, a two-period model of capacity expansion and derive the implication of varying the capacity utilization rates. In the next section the long run implication of optimal capacity expansion and its impact on efficient outputs and prices is investigated in some detail.

The term 'capacity' is often viewed as a ceiling on production or output, that is commonly referred to as the engineering definition of capacity. It has long been recognized that this definition is largely irrelevant to economics. For example, a number of empirical studies have found that the capital stock in the U.S. is idle to a significant degree for most of the time. One of the earliest studies by Foss (1963) reported an average work week of capital of only 38 hours per week. Presumably most of the idleness is either optimal or useful in the managerial discretionary behavior.

Economists view capacity rather differently. According to Winston (1974), Klein and Long (1973) full capacity describes a firm's planned or intended level of utilization; the level that reflects satisfied expectations and is built into the capital stock and embodied in the normal working schedule. Two empirical measures of capacity are commonly used in applied work in manufacturing industries. One is the U.S. Federal Reserve Board (FRB) series on capacity indexes which attempt to capture the concept of sustainable practical capacity, which is defined as the greatest level of output that a plant can maintain within the framework of a realistic work schedule, taking account of normal downtime and assuming sufficient availability of inputs to operate the machinery and equipment in place. Hence this level of output does not necessarily represent either the maximum that can be extracted from the fixed plant (as indicated by utilization rates that sometimes exceed 100 percent) or the level associated with the minimum point of the short run average cost curve. More specifically, the first step in estimating capacity indexes is to divide an industrial production index $(Q_t)$ by a utilization rate $(CU_t)$ provided by the Census Department's Survey of Plant Capacity Utilization. This yields an initial estimate of implied capacity: $IC_t = Q_t/CU_t$. However the survey is conducted every four years and firms are asked to report utilization in the fourth quarter of that year. This generally leads to cyclical variability in implicit capacity. To eliminate this cyclical volatility the second step is used to regress implied capacity $IC_t$ on capital stock $(K_t)$ and a deterministic funciton of time as

$$\ln IC_t = \ln K_t + \alpha + \sum_{i=1}^{\tau} \beta_i f_i(t); \qquad (2.4.1)$$

where $K_t$ is the year end capital stock and $f_i(t)$ is an i-th order polynomial defined on time t. The fitted values from these regressions provide baselines for the annual FRB estimates of productive capacity $(C_t)$.

A second method in estimating production capacity is to use a filter due to Hodrick and Prescott (i.e., HP filter), which decomposes a time series into a permanent and a transitory component. Hodrick and Prescott (1993) define the permanent component as including those variations which are sufficiently smooth to be consistent with slowly changing demographic and technological factors and the accumulation of capital stocks. The HP permanent component is used as a measure of capacity. Then the capacity utilization is calculated as productoin ($Q_t$) divided by the HP permanent component. In the short run, both demand and supply shocks may cause the deviations of actual output from the permanent component. Note that firms may have several options in regard to raising output above its potential level, e.g., by adding shifts, varying the production line speeds, altering the product mix or even bringing mothballed facilities back into use.

Recently Kennedy (1995) used quarterly data (1960I-1992IV) for U.S. manufacturing to regress the rate of producer price index (PPI) on both utilization rates of FRB and HP capacity variables and found the HP variable to be dominant. For example in manufacturing the HP rate coefficient is 29.10 with a t-statistic of 2.8, whereas the FRB coefficient is $-1.06$ with a t-statistic close to zero. For the disaggregated industries (two digit SIC code industries) the results are similar.

In our approach we combine the two methods above to define a series of capacity levels $CAP_{jt}$ for j-th unit at time t. This is based on two steps. In the first step we assume an additive decomposition of implied capacity into a permanent component ($IC_{jt}^P$) and a transitory component ($\zeta_{jt}$):

$$IC_{jt} = IC_{jt}^P + \zeta_{jt}$$

A filtering method (e.g., Kalman filter) is applied here to estimate the permanent component, until the random component $\zeta_{jt}$ turns out to be a white noise process. In the second step we use the data on capital stock ($K_{jt}$) and the time variable to regress $IC_{jt}^P$ on $K_{jt}$ and $f_i(t)$ as defined in (2.4.1)

$$\ln \hat{IC}_{jt} = \ln K_t + \alpha + \sum_{i=1}^{\tau} \beta_i f_{ij}(t)$$

Taking antilogs of the dependent variable we obtain the estimate $\hat{CAP}_{jt}$ of capacity. On using this capacity series $z_{jt} = \hat{CAP}_{jt}$ we set up two overall cost minimization models in the DEA framework: one involving the optimal utilization rate $\psi^*$ and the other the optimal capacity $z^*$ and optimal variable inputs $x^*$.

$$\text{Min } \theta + \xi$$

$$\text{s.t.} \quad \sum_{j=1}^{N} \lambda_j x_j \le \theta X_k ; \sum_{j=1}^{N} \mu_j z_{jt} \le \psi z_{kt}$$

$$\sum_{j=1}^{N} \lambda_j Y_j \ge Y_k ; \Sigma \lambda_j = 1 = \Sigma \mu_j \qquad (2.4.2)$$

$$\lambda_j, \mu_j \ge 0$$

and

$$\text{Min } q' x + z$$
$$\text{s.t.} \quad \sum_j \lambda_j X_j \le x ; \sum_j \mu_j z_{jt} \le z ; \sum_j \lambda_j Y_j \ge Y_k \qquad (2.4.3)$$
$$\Sigma \lambda_j = 1 = \Sigma \mu_j ; (\lambda, \mu) \ge 0$$

Here capacity z is a scalar variable, $(X_j, Y_j)$ are input output vectors for unit j and q$'$ is a row vector denoting unit costs (prices) for the variable inputs x. Denote by asterisks the optimal value of the decision variables. Then unit k is relatively inefficient in the use of capacity inputs if $\psi^* < 1.0$, whereas it is inefficient in the use of current inputs if $\theta^* < 1.0$. The optimal values $x^*, z^*$ of current and capacity input may also be compared with the actual levels $X_k, z_k$ used by unit k in order to locate efficiency gap if any. In case market price data (p) are available for the output vector y and a two-period framework is assumed, then the optimal inputs and outputs can be determined from the LP model as follows:

$$\underset{y,x,z}{\text{Max }} \pi = p_t' y - q_t' x - w_t' z + (1+r)^{-1} w_{t+1}' \hat{z}$$
$$\text{s.t.} \quad X\lambda \le x ; Y\lambda \ge y ; \mu' z_t \le z ; \mu' \hat{z}_t \ge \hat{z} \qquad (2.4.4)$$
$$\lambda' e = 1 = \mu' e ; \lambda \ge 0, \mu \ge 0$$

Here $z_t$ is a vector of durable inputs puchased at the the beginning of period t at prices $w_t$, $\hat{z}_t$ is a vector of depreciated durable inputs that will be available to the firm at the beginning of the subsequent period, $w_{t+1}$ is the vector of durable input prices that the firm anticipates will prevail during period t+1, and r is an appropriate discount rate exogenously given. Here the capacity-related inputs are the durable inputs and their unit costs are the input prices. With observed values $(X, Y, z_t, \hat{z}_t)$ of inputs and outputs the firm could now determine the optimal iputs and outputs $(x^*, y^*, z^*, \hat{z}^*)$. We note however some basic differences of this formulation from the traditional DEA models. First of all, the vector of spot prices $w_{t+1}$ is not observed at time t and hence the producer's anticipation of future price is needed. In this sense this model yields anticipated or expected efficiency. Since the anticipated prices are uncertain, the firm's attitude towards uncertainty must be modeled. This is the framework where the rational expectations (RE) hypothesis may be introduced. Secondly, the durable inputs are used here to approximate the stock of capacity inputs, but for certain stocks like natural resources and goods inventories there may be no natural market prices. Finally, the relevant discount rate r must be common to all the firms and also known. In the static DEA models these basic questions are not addressed at all.

In case of stochastic demand $d_t$ the DEA model (2.4.4) can be transformed as:

$$\text{Max } E \pi = \left[ p'_t \min(y, d_t) - q'_t x + (1+r)^{-1} \hat{z} - w'_t z \right]$$

s.t.        the same constraints as (2.4.4)

where E denotes expectation. On using the Lagrangean expression:

$$L = E\pi + \beta'(x - X\lambda) + \alpha'(Y\lambda - y) + \gamma'(z - \mu'z_t) + \delta'(\mu'\hat{z}_t - \hat{z})$$
$$\beta_0(1 - \lambda'e) + \gamma_0(1 - \mu'e)$$

We must have for the efficient unit:

$$p_t(1 - F(y^*)) - \alpha^* = 0$$
$$\gamma^* = w_t; \beta^* = q_t; \delta^* = (1+r)^{-1}$$
$$Y'\alpha^* = X'\beta^* + \beta_0^* e; \gamma^{*'} z_t = \delta^{*'} \hat{z}_t - \gamma_0^* e$$

This shows that the unit exhibits output inefficiency if $\sum\limits_{j=1}^{N} Y_j\lambda_j^* > y^*$, input efficiency if $\Sigma X_j\lambda_j^* < x^*$, capacity inefficiency if $\mu'^* z_t < z$ or $\mu^{*'} z_t > \hat{z}^*$. Clearly there are five sources of inefficiency in this framework: the input, output, capacity and inefficiency due to market demand uncertainty. The theory of organizational slack deals specifically with the demand and capacity oriented sources of inefficiency which may apparently inflate the marginal costs.

In case we have a time horizon it is simpler to introduce investment variables denoted by a vactor $I_t$ and rewrite the long run profit function as

$$\text{Max } E \left\{ \sum\limits_{t=0}^{\infty} (1+r)^{-t} [p'_t \min(y_t, d_t) - q'_t x_t - \rho'_t I_t - w'_t z_t] \right\}$$

s.t.        $X_t\lambda \leq x_t; Y_t\lambda_t \geq y_t; I_t \leq (1 + \delta_0)z_{t+1} + z_t$
          $\lambda'_t e = 1; \lambda_t \geq 0$

where investment is constrained by changes in capacity inputs with $\delta_0$ denoting fixed rates of depreciation. The theory of adjustment costs which relates current production to capital stock and investment in new capital along with the variable inputs is implicit in this formulation and its implications have been discussed by

Artus and Muet (1990) in an empirical framework and by Sengupta (1995b) in the DEA framework.

For public sector enterprises however the market prices of output are generally unavailable and the profit maximization objective does not apply, since these are not for profit organizations. Hence in this case we may restrict ourselves to the cost frontier alone and use the theory of adjustment costs to develop a model of capacity utilization. Consider the production function

$$y = f(v,x)$$

of a firm, which produces a single output y by means of the vector v of variable inputs and the vector x of service flows from the quasi-fixed inputs (i.e., these inputs are fixed in the short run but variable in the long run). Since the production function may exhibit increasing returns to scale, the usual profit maximization principle may not yield determinate results. Hence we adopt the cost minimizatio model, where in the short run the firm minimizes variable costs $q' v$ in the short run subject to the production constraint $y \leq f(v,x)$ where x is fixed. This yields the short run cost function $C_v = g(y,q,x)$. Denoting by w the vector of rental prices for the quasi-fixed inputs, the total cost $C = C_v + C_x$ may be defined with $w' x = C_x$ as th fixed cost. Capacity output $\hat{y}$ is now defined by that level of output for which total cost C above is minimized, i.e.,

$$\hat{y} = h(q, x, w) \qquad (2.4.5)$$

with the associated cost function for capacity output

$$\hat{C} = G(q, w, \hat{y}) \qquad (2.4.6)$$

Two implications of this concept of optimal capacity output must be noted. One is that the capacity output $\hat{y}$ may be viewed as a point of tangency between the short and the long run total cost curves. Secondly, one can now define the rate of capacity utilization as $u = y / \hat{y}$ where $0 \leq u \leq 1$. Morrison and Berndt (1981) used this type of a dynamic cost function model with a single quasi-fixed input called capital to estimate the patterns of capacity utilization of U.S. manufacturing over the period 1958-77 by using a regression model. One can also use a DEA model to specify a cost frontier as follows:

$$\text{Min } s_k + \hat{s}_k$$
$$\text{s.t.} \quad a_0 + \sum_i a_i q_{ij} + \sum_i b_i x_{ij} + dy_j + s_j = C_j$$
$$a_0 + \sum_i \alpha_i q_{ij} + \sum_i h_i w_{ij} + \delta \hat{y}_j + \hat{s}_j = \hat{C}_j$$
$$j = 1, 2, \ldots, N$$

where $C_j$ and $\hat{C}_j$ are observed short run and long run costs for unit j and the observed data consist of input and output prices and the two outputs. If unit k is efficient then we must have $s_k^*$ and $\hat{s}_k^*$ to be zero implying full capacity utilization.

If the short and long run cost components can be separately obtained as $C_{ij}$ and $\hat{C}_{ij}$ then these could be used more directly to characterize DEA efficiency as follows:

$$\text{Min } \varepsilon + \zeta$$
$$\text{s.t.} \quad \sum_{j=1}^{N} C_{ij}\lambda_j \leq \varepsilon C_{ik} \,; \Sigma\lambda_j = 1, \lambda_j \geq 0$$
$$\sum_{j=1}^{N} \hat{C}_{ij}\varpi_j \leq \zeta\hat{C}_{ik} \,; \Sigma\mu_j = 1, \mu_j \geq 0$$
$$j=1,2,\ldots,N$$

In this framework unit k is efficient in the short run if $\varepsilon^* = 1.0$, but not efficient in the long run if $\zeta^*$ is less than one. The fact that some inputs are fixed in the short run makes it clear that the rate of capacity utilization may influence short and long run costs differently.

From an economic viewpoint the most important source of excess capacity is due to a fall in market demand, i.e., demand uncertainty and the existence of excess capacity tends to inflate the short and the long run costs of output. For public sector enterprises the competitive market pressure is very weak, hence the probability of incurring dead weight losses and hence inefficiency due to organizational slack is much higher.

## 2.5. Quality and Efficiency

Modelling efficiency measurement in a nonparametric way has recently found widespread applications in public sector organizations such as public school and colleges, public clinics and hospitals and private sector enterprises such as commercial banks, investment firms and commercial airlines, see e.g., Fried et al. (1993), Charnes et al. (1994) and Sengupta (1995, 1998a). However these applications have neglected to incorporate the quality improvements in inputs and outputs, which usually accompany productivity improvements in manufacturing and service industries. There has been a dramatic surge in the focus on quality issues by U.S. firms in the recent decade. Reasons are several. First of all, increased competition faced by U.S. firms especially from Japanese competitors has generated increased emphasis on quality in U.S. manufacturing. For example

Garvin (1998) noted that in the automobile industry in 1979, Toyota averaged 0.71 defects per vehicle shipped, while Ford averaged 3.70 defects. This in turn led to U.S. automobile firms' improving the quality of their products. Secondly the quality-based learning curves play a very crucial role in reducing unit costs in the recent growth industries such as microelectronics, semiconductors and telecommunications. Unless such cost improvements are introduced, the efficiency estimates by the DEA models tend to be seriously biased. Finally, the learning by doing models in new growth theory emphasize the effect that learning improves the efficiency with which firms can perform quality-related activities. Arrow (1962) emphasized the scale effect of cumulated gross investment as the basis of learning by doing, while Lucas (1993), Norsworthy and Jang (1992) and Jovanovic (1997) laid stress on the cumulative experience in output production as the basis of productivity and quality improvement. Since adaptivity is the core of the learning process, it necessarily entails a dynamic process of adjustment over time. Thus the intertemporal dynamic aspect of quality improvement plays a critical role in the quality-based efficiency frontier.

Our object in this section is to generalize the DEA-type efficiency measure in productivity analysis by introducing quality-based learning by doing and thereby evaluate the dynamic implication of a quality-based cost efficiency frontier, where it is assumed that quality-based cumulative experience affects quality control costs.

Although quality is defined in many different forms in the operations management literature, two concepts of quality stand out. One defines it as the degree of conformance to design specification. This corresponds to the view of the quality control technicians. The second view considers quality of the design itself, where it is suggested for firms' decision making that product (design) quality should be as high as possible, since its benefits in terms of demand expansion far outweigh the costs of investment in quality improvement. This second view has been emphasized strongly by Jovanovic (1997) and others in respect of learning by doing models in new growth theory in economics. In this study we will use both the concepts of quality, e.g., both design and conformance quality characteristics, noting however that the second concept is more cost oriented, while the first is demand oriented.

Consider now a standard input oriented DEA model for testing the Pareto efficiency of a reference firm or decision making unit h ($DMU_h$) in a cluster of N units, where each $DMU_j$ products s outputs ($y_{rj}$) with m inputs ($x_{ij}$)

$$\text{Min } \theta \text{ subject to } \sum_{j=1}^{N} X_j \lambda_j \leq \theta X_h$$

$$\sum_{j=1}^{N} Y_j \lambda_j \geq Y_h; \sum_{j=1}^{N} \lambda_j = 1; \lambda_j \geq 0 \qquad (2.5.1)$$

Here $X_j$ and $Y_j$ are the observed input and output vectors for each $DMU_j$ where $j=1,2,...,N$ and $\lambda_j$ are the nonnegative weights to be optimally determined along with the scalar variable $\theta$. Let $\lambda^*$ and $\theta^*$ denote the optimal solutions of the linear programming (LP) model (2.5.1) with all slacks zero. Then the reference unit $DMU_h$ is said to be *technically efficient* (i.e., production efficiency) if $\theta^* = 1.0$. If however $\theta^*$ is positive but less than unity, then it is not technically efficient, since it uses excess inputs measured by $(1-\theta^*)x_{ih}$. Thus in case of technical inefficiency, the first constraint shows that a linear combination of other firms in the industry does better in using less inputs, while the second constraint, if it holds with inequality shows that a linear combination of other firms does better also in terms of producing more output than the reference firm.

While technical efficiency (TE) does not depend on the input prices at all, allocative efficiency (AE) measures a firm's succes sin choosing an optimal set of inputs with a given set of input and output prices. Overall efficiency ($OE_j$) of a unit j combines both technical ($TE_j$) or production efficiency and the allocative ($AE_j$) or price efficiency as follows: $OE_j = TE_j \times AE_j$. To measure the overall efficiency of a reference unit or $DMU_h$ one solves the cost minimization LP model

$$\text{Min } C = c' x$$
$$\text{s.t.} \quad X\lambda \leq x; Y\lambda \geq Y_h; \lambda'e = 1; \lambda \geq 0 \tag{2.5.2}$$

where e is a column vector with each element unity, prime denotes transpose, $c = (c_j)$ is a unit cost (or price) vector for the inputs $x = (x_i)$ and $X = (X_j)$, $Y = (Y_j)$ are appropriate matrices of observed inputs and outputs. Denote optimal values by asterisks. Then, technical efficiency of the reference unit $DMU_h$ is $TE_h = \theta^*$ as before, overall efficiency is $OE_h = C_h^*$ where $C_h^*$ is the optimal value of the objective function C in (2.5.2) and hence the allocative efficiency is $AE_h = C_h^*/(q^*C_h)$. When the competitive output prices $p = (p_r)$ are available and the production costs are available in the form of output costs, the above model (2.5.2) can be rewritten as

$$\text{Max } z = p' y - v' y - \theta = \sum_{r=1}^{s} (p_r y_r - v_r y_r) - v_0 - \theta$$
$$\text{s.t.} \quad A\lambda \leq \theta X_h; Y\lambda \geq y; \lambda''e = 1; \lambda \geq 0 \tag{2.5.3}$$

where $v_0 + \sum_{r=1}^{s} v_r y_r$ is the total cost of the output vector y. The efficient unit is now characterized by the following conditions

$$p_r = v_r + \alpha_r^*; \beta^{*'} X_h = 1$$
$$\alpha^{*'} Y_j = \beta^{*'} X_j + \alpha_0^*; \quad j=1,2,\ldots,N \tag{2.5.4}$$

where the Lagrangean function is

$$L = \sum_{r=1}^{s} (p_r - v_r)y_r - \theta - v_0 + \beta'\left(\theta X_k - \sum_{j=1}^{N} X_j\lambda_j\right)$$
$$+ \alpha'\left(\sum_{j=1}^{N} Y_j\lambda_j - y\right) + \alpha_0(1 - \lambda'e)$$

Now consider the quality variable $q_{rj}$ for each output $r=1,2,\ldots,s$ for each DMU$_j$ and rewrite the total cost function for each unit as

$$C(y,q) = v_0 + \sum_{r=1}^{s} \left[(v_r + \varepsilon_r q_r)y_r + f_r + (1/2)g_r q_r^2\right] \tag{2.5.5}$$

The transformed model is then

$$\text{Max } z = p'y - C(y,q) - q \tag{2.5.6}$$
$$\text{s.t.} \quad \sum_{j=1}^{N} Q_j\lambda_j \le q \text{ and the constraints of } (2.5.3)$$

where $Q_j = (q_{rj})$ is the observed quality vector for unit j. Here quality is viewed in terms of reliability and measured by the expenditure on it. The decision vectors here are the outputs, the quality expenditures and the weight variable $\theta$. The form of the cost function (2.5.5) assumes that the quality level selected by a firm affects total costs in three ways:

(i)    investment in a quality improvement program increases fixed productoin costs. Thus the fixed costs $\sum_{r} (f_r + (1/2)g_r q_r^2)$ are increasing ($g_r > 0$) and convex in the quality level $q_r$;

(ii)    the quality level also influences the unit production cost, e.g., for any given quality level $q_r > 0$ selected by the firm, the unit variable cost increases (decreases) by $|\varepsilon_r q_r|$, where $\varepsilon_r$ is either positive or negative. A negative value of $\varepsilon_r = -\hat{\varepsilon}_r, \hat{\varepsilon}_r > 0$ implies that the production costs actually decline when quality is improved. A positive value shows an increase in production costs as in the traditional quality control literature when quality in the sense of reducing the frequency of defects and unreliability of the products is improved;

(iii)    the quality expenditure level in the constraint $Q\lambda \leq q$ has an optimal shadow price $\gamma^* = (\gamma_r^*)$, which has an impact on the optimal output levels $y^*$ and hence on the production cost frontier. By assuming the quality constraint as an equality $Q\lambda = q$, the shadow prices $\gamma_r^*$ may be made flexible in the sense that it may take on any sign, i.e., positive, negative or zero. Then the production frontier becomes flexible in the sense that the quantity of output may be correlated with quality either positively or negatively. The quality constraint equality $Q\lambda = q$ may be viewed as a goal or target.

For the quadratic model (2.5.6) the efficient DMU must satisfy the following necessary conditions which are also sufficient because of the strong convexity assumption for the cost function $C(y,q)$

$$p_r = v_r - \hat{\varepsilon}_r q_r^* + \alpha_r^*; \beta^{*\prime} X_h = 1$$
$$q_r^* = (1/g_r)[\hat{\varepsilon}_r y_r^* - \gamma_r^*] \tag{2.5.7}$$
$$\alpha^{*\prime} Y_j = \beta^{*\prime} X_j + \gamma^{*\prime} Q_j + a_0^*$$

On comparing the results from (2.5.7) and (2.5.4) one obtains several interesting economic implications. First of all, if the production cost declines (i.e., $\hat{\varepsilon}_r > 0$) as quality improves, then the equilibrium price falls; in this case the higher the optimal output, the higher the quality level. Secondly, the higher optimal output has a positive marginal impact on quality. Also, if we drop the term $p' y$ from the objective function, one can easily compute the overall efficiency of $DMU_h$ as $OE_h = C(y^*,q^*)/C(y,q)$, whereas $TE_h = \theta^*$ and $AE_h = C_h^* /(\theta * C)$, where $C_h^* = C(y^*,q^*)$ and $C = C(y,q)$. This shows that optimal quality is closely connected with optimal quantity of output and the associated cost frontier $C^*$. It is clear that if $\varepsilon_r > 0$ then higher quality entails lower optimal output and also higher prices in equilibrium. Finally, the production frontier relation in (2.5.7) shows that improved quality augments (decreases) output if the shadow prices $\gamma_r^*$ are negative (positive). Since the efficiency model (2.5.6) is a quadratic program, the optimal solutions permit more substitutions between quality and quantity and hence between quality and efficiency. Thus the quality-adjusted DEA model of efficiency incorporates three additional sources of efficiency not found in conventional DEA models, e.g., (1) complementarity effect of quality when it reduces unit production costs, the increasing returns to scale due to learning by doing incorporated through the parameter $\varepsilon_r < 0$ and, the trade-off of quality and quantity along the production frontier.

## A.. Quality improvement and learning over time

Quality improvement involves a learning process over time, where the learning curve approach suggests that effort devoted to improving the quality of manufactured products may reduce unit production costs. Here experience is usually modeled in terms of cumulative output, where output is adjusted for quality. Thus the cumulative experience embodied in output type r may be written as

$$\tilde{y}_r(t) = \tilde{y}_r(0) + \int_0^t y_r(s)q_r(s)ds \qquad (2.5.8)$$

The cost function in (2.5.5) is now written as

$$C(y, \tilde{y}, q) = v_0 + \sum_{r=1}^s \left\{ [v_r(\tilde{y}) + \varepsilon_r q_r] y_r + f_r + (1/2) g_r q_r^2 \right\} \qquad (2.5.9)$$

where we assume that quality-based learning benefits accrue in direct production costs, where the marginal cost $v_r(\tilde{y}_r)$ declines as cumulative output rises. A transformed DEA model may now be set up which minimizes the discounted cost function as follows:

$$\text{Min } J = \int_0^T \left\{ e^{-\rho t} C(y, \tilde{y}, q) dt + \theta(t) \right\}$$

s.t.        the same constraint of model (2.5.6)                     (2.5.10)

Here $\rho$ is the exogenous discount rate and the decision variable $y(t)$, $\tilde{y}(t)$, $q(t)$, $\lambda(t)$ and $\theta(t)$ are all time varying within a horizon $[0,T]$. Clearly this minimization problem is a calculus of variations problem subject to inequality constraints. Let L be the Lagrangean function

$$L = e^{-\rho t}[-C(y, \tilde{y}, q) - \theta + \beta'(\theta X_h - X\lambda)$$
$$+ \gamma'(-Q\lambda + q) + \alpha'(Y\lambda - y)$$
$$+ \alpha_0(1 - \lambda'e) + \sum_{r=1}^s \mu_r(y_r q_r - \dot{\tilde{y}}_r)]$$

Then the necessary conditions of optimality for r=1,2,...,s can be written as

$$\dot{\mu}_r = \rho\mu_r + MC_r(\tilde{y}_r)y_r; \mu_r(T)e^{-\rho T} = 0$$
$$(\mu_r - \varepsilon_r)y_r \leq g_r q_r + \gamma_r; (\mu_r - \varepsilon_r)q_r \leq v_r + \alpha_r \qquad (2.5.11)$$

$$\beta'X_h = 1 \text{ and } \alpha'Y_j \le \alpha_0 + \beta'X_j + \gamma'Q_j; j=1,2,\dots,N$$

Since $q_r = \dot{\tilde{y}}_r / y_r$ this quality-based learning model must also satisfy the following dynamic conditions

$$\dot{\tilde{q}}_r = (1/g_r)[y_r MC(\tilde{y}_r) + \rho(\varepsilon_r + g_r \tilde{q}_r) + (\rho - G_r)(\gamma_r/y_r)] \tag{2.5.12}$$

where $MC(\tilde{y}_r) = \partial v_r(\tilde{y}_r)/\partial \tilde{y}_r, \tilde{q}_r = q_r/y_r$ and $G_r$ denotes the growth rate of output $\dot{y}_r/y_r$. Clearly if the marginal cost $MC(\tilde{y}_r)$ declines, i.e., $MC(\tilde{y}_r) = -m_r$, $m_r > 0$ and there is no discounting then the time rate of change of quality per unit of output is negative. With discounting $\dot{\tilde{q}}_r$ is positive or negative according as

$$\rho(\varepsilon_r + q_r \tilde{q}_r) + \frac{\gamma_r \rho}{y_r} \mathop{\gtrless}\limits G_r + m_r y_r$$

In the steady state as T goes to infinity one obtains the relation

$$q_r = (1/g_r)[(m_r/\rho)y_r^2 - \varepsilon_r y_r] \tag{2.5.13}$$

This shows that in the steady state the quality level $q_r$ is a strictly convex function of the quantity of output. Thus there exists an output level $y_r^0$ at which the equality reaches the minimum level $q_r^0$

$$y_r^0 = (\varepsilon_r \rho)/(2m_r)$$

As $y_r$ increases above $y_r^0$, the quality level $q_r$ rises above $q_r^0$. If however there is no cost reducing effect of cumulative output, i.e., $m_r = 0$, then $q_r$ falls as $y_r$ rises in the steady state.

Thus the dynamic quality frontier specified in equation (2.5.12) has several interesting implications. First of all, it shows that it has both decreasing and decreasing phases over time depending on the steady state condition (2.5.13) and it follows that

$$\partial q_r/\partial m_r > 0, \partial q_r/\partial \rho < 0, \partial q_r/\partial \varepsilon_r < 0$$

Spence (1981) analyzed some of these issues relating to the impact of quality on cost efficiency. Secondly, the static efficiency of a DMU$_h$ would not necessarily be dynamically efficient unless it satisfies the dynamic optimality conditions in (2.5.11). Thirdly, it is clear from the optimality condition in (2.5.11) that the output $(y_r)$ and quality $(q_r)$ are positively (negatively) correlated if $\mu_r$ exceeds (falls short of) $\varepsilon_r$. Thus the adjoint variable $\mu_r$ plays a critical role in the trade off between quality and quantity, where the change in $\mu_r(t)$ is determined by the differential equation

$$\dot{\mu}_r(t) = \rho\mu_r - m_r y_r(t) = \rho\mu_r - m_r \dot{\hat{y}}_r / q_r \qquad (2.5.15)$$

This shows that as quality $q_r(t)$ rises, $\dot{\mu}_r(t)$ rises, whereas if $\dot{\hat{y}}_r$ rises, $\dot{\mu}_r(t)$ falls. In the steady state when $\dot{\mu}_r(t) = 0$ one obtains

$$y_r = \rho\mu_r/m_r$$

Finally, we note that the static production frontier in (2.5.11) for $DMU_h$

$$\alpha' Y_h = \alpha_0 + \beta' X_h + \gamma' Q_h$$

cannot be fully specified unless the shadow price vector $\alpha = (\alpha_r)$ satisfies the relation $(\mu_r - \varepsilon_r)q_r = v_r + \alpha_r$, where $\alpha_r$ depends on the variable $\mu_r(t)$ which changes over time in terms of the differential equation (2.5.15), which is often called the perfect foresight condition. Thus static efficiency affords only a partial and often times biased view of efficiency over time.

## B. Quality and market competition

Recent surge in interest on quality management has increasingly attributed the emphasis on quality by U.S. firms to increased competition faced by them. Thus Banker et al. (1998) has modeled how quality improvement in firms occurs through increased degrees of competition. By expanding the market through advertising and international trade the firms can achieve strong economies of scale thereby reducing production costs and increasing productivity. The standard Farrell and DEA models have neglected this aspect altogether. We adopt here a static and a dynamic model to specify quality improvement and efficiency. We assume quality $(q_r)$ to be an attribute of output $(y_r)$ so that in quality units output can be written as $\hat{y}_r = y_r q_r$. It is further assumed that the equality vector $Q_j = (q_{rj})$ is observable for each DMU and each DMU seeks to attain the highest quality level possible. The static DEA model may then be written as

$$\text{Max } \pi = \sum_{r=1}^{s} \left[ p_r y_r - (v_r - \varepsilon_r q_r)y_r - (1/2)g_r q_r^2 - f_r \right] - \theta$$

$$\text{s.t.} \quad \sum_{j=1}^{N} X_j \lambda_j \leq \theta X_h; \sum_{j=1}^{N} \hat{Y}_j \lambda_j \geq \hat{y}; \sum_{j=1}^{N} Q_j \lambda_j \geq q \qquad (2.5.16)$$

$$\lambda' e = 1, \lambda \geq 0$$

It is assumed here that output prices $(p_r)$ are given and $\varepsilon_r$ is a positive constant specifying cost reduction due to quality improvement. Here the decision variables are output $(y_r)$, quality attributes $(q_r)$, $\theta$ and $\lambda_j$. The necessary conditions for optimality can then be specified as

$$\beta' X_h = 1; p_r \leq v_r + (\alpha_r - \varepsilon_r) q_r$$
$$g_r q_r \geq (\varepsilon_r - \alpha_r) y_r - \gamma_r$$

and

$$\alpha' \hat{Y}_j \leq \beta' X_j - \gamma' Q_j + \alpha_0$$

where the Lagrangean function is

$$L = \pi + \beta'(\theta X_h - \Sigma X_j \lambda_j) + \alpha'(\Sigma \hat{Y}_j \lambda_j - \hat{y})$$
$$+ \gamma'(\Sigma Q_j \lambda_j - q) + \alpha_0 (1 - \lambda' e)$$

If $DMU_h$ is efficient then one must have $\theta = 1.0$ and

$$q_r = (1 / g_r)[(\varepsilon_r - \alpha_r) y_r - \gamma_r]$$

which shows that

$$\partial q_r / \partial y_r \overset{\geq}{<} 0 \text{ as } \varepsilon_r \overset{\geq}{<} \alpha_r \qquad (2.5.18)$$

If $\varepsilon_r$ is zero then $\partial q_r / \partial y_r$ is always negative, since $\alpha_r$ is positive for the efficient unit. One may compare this quality-adjusted model with the traditional model where the quality attribute is dropped and we obtain for the efficient DMU the following relations

$$\beta' X_h = 1; p_r = v_r + \alpha_r$$
$$\alpha' Y_j = \beta' X_j + \alpha_0 \qquad (2.5.19)$$

Clearly the production frontier does not incorporate any effect of quality improvement or decline and the optimal price does not fall when quality improves. In recent times the technology-intensive industries such as electronics, semiconductors and telecommunications the trend in price declines has been a pervasive feature as it has been empirically shown by Norsworthy and Jang (1992).

Now consider a dynamic version of the qulaity adjusted output model. To simplify notation we now denote by $y = (y_r(t))$ the quality-adjusted output denoted before by $\hat{y}_r = y_r q_r$. We assume the cost function approach as before and since the market is assumed to be competitive, each firm or DMU has the objective of minimizing total expected costs, where the latter comprise both production costs and the expected costs of inventory and shortages when outputs deviate from expected demand. This model then shows directly how advertising can improve sales and how higher sales can reduce the production costs of quality adjusted output $y_r(t)$. Womer (1979) has discussed managerial models where higher cumulative sales tend to reduce production and quality costs.

Let $z = (z_r(t))$ and $S = (S_r(t))$ be cumulative output and demand (sales),where the latter is a random vector. The current output vector is $y = (y_r(t)) = \dot{z}(t)$, where $Y_j$ denotes the output vector of $DMU_j$. Two types of costs are incurred: the expected production cost $c(\dot{z}, S)$ and expected inventory costs $(w/2) E \sum_{r=1}^{s} (z_r - S_r)^2$ which are assumed to be proportional to the gap between cumulative production and sales. The production cost function $c(\dot{z}, S)$ is assumed to be strictly convex in current output $y = \dot{z}$ but the marginal cost $\partial c(y,S)/\partial S$ declines as cumulative sales increase. This is the scale effect. Since current output $y$ is quality adjusted, the decline in marginal cost may be due in part to the quality improvement as in (2.5.16) before.

The general dynamic efficiency model may now be specified for testing the efficiency of $DMU_h$

$$\text{Min } J = E_t \int_0^\infty e^{-\rho t} \left[ \sum_{r=1}^{s} \left\{ c_r(y_r, S_r) + (w/2)(z_r - S_r)^2 \right\} \right] dt + \theta$$

$$\text{s.t.} \quad \sum_{j=1}^{N} Y_j \lambda_j \geq \dot{z}; \sum_{j=1}^{N} X_j \lambda_j \leq \theta X_h; \lambda'e = 1; \lambda \geq 0 \qquad (2.5.20)$$

When cumulative sales can be directly expanded by promotional or advertising expenditure $(A_r)$, we can assume

$$S_r = S_{0r} + f_r A_r$$

and adjoin a set of constraints

$$\sum_{j=1}^{N} A_{rj} \lambda_j \leq A_r; \quad r=1,2,\ldots,s \qquad (2.5.21)$$

and a cost component $\sum\limits_{r=1}^{s} c_r(A_r)$ in the objective function. The vector $A = (A_r)$ of advertisement inputs may then be treated as a control or policy variable.

The nonlinear model (2.5.20) may now be directly solved by using the Euler-Lagrange necessary conditions, which are also sufficient due to the assumption of convexity of the cost functions. The necessary conditions for dynamic efficiency can be specified into two parts: a static part and a dynamic part as follows:

$$\alpha'Y_j \leq \beta'X_j + \alpha_0; \beta'X_h = 1 \tag{2.5.22}$$
$$\alpha \geq 0; \beta \geq 0; \alpha_0 \text{ free in sign}$$

and

$$\rho(MC_r(y_r) + \alpha_r) - \frac{\partial MC_r(y_r)}{\partial y_r}\dot{y}_r - \dot{\alpha}_r + w(z_r - \bar{S}_r) \leq 0;$$

$$r = 1, 2, \ldots, s \tag{2.5.23}$$

where $\bar{S}_r$ is the expected value of $S_r$ and $MC_r(y_r)$ is the marginal cost of $c_r(\dot{z}_r, S_r)$ with respect to $y_r$. Assume a specific form of the cost function in order to get simple results, e.g., $MC_r(\dot{z}_r, S_r) = m_r y_r / S_r$. Then if $DMU_h$ is both statically and dynamically efficient, we must have the following conditions satisfied

$$\alpha^{*'} Y_h = \beta^{*'} X_h + \alpha_0^*; \beta^{*'} X_h = 1 \tag{2.5.24}$$

$$\frac{m_r}{S_r}\ddot{z}_r^* - \frac{\rho m_r}{S_r}\dot{z}_r^* - wz_r^* = \rho\alpha_r^* - w\bar{S}_r - \dot{\alpha}_r^* \tag{2.5.25}$$

$$\lim_{t \to \infty} e^{-\rho t}\left(\frac{m_r \dot{z}_r^*}{S_r} + \alpha_r^*\right) = 0$$

Clearly in the steady state we would have

$$z_r^* = \bar{S}_r - \rho\alpha_r^* / w \tag{2.5.26}$$

This implies that cumulative output along the efficient path would rise, when $\bar{S}_r$ rises or $\rho$ (i.e., $\alpha_r^*$) falls. The second order differential equation in (2.5.25) has the characteristic equation

$$(m_r / S_r)\mu^2 - (\rho m_r / S_r)\mu - w = 0$$

with two real roots $\mu_1, \mu_2$ having opposite signs $\mu_1 > \rho > 0$ and $\mu_2 < 0$. Clearly the unstable root $\mu_1$ leads to an unbounded value of $z_r^*(t)$ where

$$z_r^*(t) = A_1 e^{\mu_1 t} + A_2 e^{\mu_2 t} + z_r^*$$

Hence by the transversality condition $A_1$ must be set equal to zero. Thus one obtains the convergence that $z_r^*(t)$ tends to the steady state level $z_r^*$ in (2.5.26) as $t \to \infty$. This convergence has several economic implications. First of all, learning involves here the adaptive process of convergence to the steady state. Second, if we adjoin addiitonal control variables such as $A_r$ in (2.5.21), then the timal value of $A_r$ would satisfy the following condition

$$A_r^* = (1/f_r)\left[z_r^* - S_{0r} + \gamma_r^*(wf_r)^{-1}\right]$$

which shows that cumulative sales and advertisement expenditures are highly positively correlated and higher values of w lead to lower $A_r^*$. Finally, the dynamic efficiency path (2.5.25) involves both the current value of $\alpha_r^*(t)$ and its time rate of change $\dot{\alpha}_r^*(t)$. The static efficiency condition (2.5.22) in standard Farrell or DEA models ignores the time rate of change of $\alpha_r^*(t)$.

Furthermore, if we normalize $\overline{S}_r$ in terms of the standard deviation $\sigma_r$ of $S_r$ then the steady state equation (2.5.26) may be written as

$$z_r^* = (\overline{S}_r / \sigma_r) - \rho\alpha_r^* / w$$

which would imply that higher demand fluctuations would entail lower cumulative output on the efficiency frontier.

Thus the impact of demand and its fluctuations on the efficient output can be analyzed both in the short and the long run. Thus we have the result that when the relative efficiency concept of Farrell and DEA models is extended to dynamic frameworks, the phenomenon of learning by doing acquires an important role. Cumulative experience through inputs or outputs tends to reduce production costs and the learning parameters in the form of quality improvements become as important as the productivity parameters.

# 3. Efficiency Dynamics

New growth theory in macroeconomics has emphasized very strongly the dynamic changes in industrial productivity over time. Solow (1956) in his classic contribution to economic growth assumed a loglinear production function Y(t) with two inputs: labor (L) and capital (K) under constant returns to scale

$$Y(t) = A(t) K^{\alpha} L^{1-\alpha} \tag{3.1}$$

and introduced the competitive market assumption that the two inputs are paid the value of their marginal products. This yields the neutral technological progress function as

$$\dot{A}/A(t) = \dot{Y}/Y(t) - w_K (\dot{K}/K) - w_L (\dot{L}/L) \tag{3.2}$$

where $w_K$ and $w_L$ are the shares in total costs of production of the two inputs and dot over a variable denotes its time derivative. The term A(t) in the production function (3.1) is called Hicks-neutral technical change, which indicates shifts in production function (or the production frontier if it refers to an efficient production function) which leave marginal rate of input substitution unchanged. The term $(\dot{A}/A)$ is frequently termed total factor proeductivity (TFP) growth, also as Solow residual.

In a more general framework with several inputs and no assumption about the factor market equilibrium one could derive the TFP growth equation as

$$\dot{A}/A = \dot{Y}/Y - \sum_{i=1}^{m} \beta_i(\dot{x}_i/x_i) \tag{3.3}$$

where $x_i$ denotes i-th input with its productin elasticity $\beta_i$. if the sum of the coefficient $\beta_i$ exceeds (equals) one, we have increasing (constant) returns to scale. Two points are most important in this type of dynamics. One is that it measures long run rate of growth of output as a linear function of the growth rates of different inputs, thus involving the TFP growth due to technical progress. This is quite different from short run increases in output which do not involve rates of growth. The standard DEA model involves changes (increases) in the level of output and hence *level efficiency*, whereas the TFP growth models (3.2) and (3.3) involve *growth efficiency*. Clearly if we have panel data for several years, both these efficiencies can be evaluated by the modified DEA model. Secondly, the assumption of market equilibrium is very crucial in the derivation such as (3.2). Thus allocative efficiency in growth underlies the formulation (3.2), whereas technical efficiency is characterized by (3.3). In the case when the market is characterized by imperfect competition one could extend such formulations.

Besides technology two other factors have been strongly emphasized in new growth theoy, which stimulate long run growth. One is the knowledge capital which has significant spillover effects and externalities across firms, industries and even different countries. The skill aspect of human capital, R&D (research and development) inputs and new vintages of software technology used in the production process are some of the proxy variables which can capture the role of knowledge capital. The second factor behind the growth of new industires is the information technology embodied in flexible manufacturing systems (FMS), which differs from the fixed manufacturing systems in terms of adaptivity to new market conditions and flexibility for new innovations. Both these aspects of efficiency dynamics are important for the DEA models of efficiency, if they are to be applied to modern technology-intensive firms such as microelectronics, software companies and telecommunications.

## 3.1    Sources of Productivity Growth

Consider a two input model of single output $(y_j)$ with labor $(x_{1j})$ and capital $(x_{2j})$ inputs, which characterizes the production efficiency frontier by a free-disposal convex hull

$$\text{Min } \theta \text{ subject to } \sum_{j=1}^{N} x_{ij}\lambda_j \leq \theta x_{ik} \text{ ; } i=1,2$$

$$\sum_{j=1}^{N} y_j\lambda_j \geq y_k ; \Sigma\lambda_j = 1, \lambda_j \geq 0 \qquad (3.1.1)$$

This is the input-oriented version of the DEA model. In the output-oriented case one obtains

$$\text{Max } \phi$$

$$\text{s.t.} \quad \sum_{j=1}^{N} X_j\lambda_j \leq X_k ; \sum_{j=1}^{N} y_j\lambda_j \geq \phi y_k \qquad (3.1.2)$$

$$\Sigma\lambda_j = 1, \lambda_j \geq 0 \text{ ; } j=1,2,\ldots,N$$

where $X_j$ is the input vector for unit or $DMU_j$. If we have constant returns to scale, then one can specify a combined model as

$$\text{Max } \phi-\theta$$
$$\text{s.t.} \quad \Sigma X_j \lambda_j \leq \theta X_k; \Sigma y_j \lambda_j \geq \phi y_k$$
$$\Sigma \lambda_j = 1, \lambda_j \geq 0 \tag{3.1.3}$$

The dual problem for (3.1.1) can be easily derived as

$$\text{Max } \beta_0 - \alpha \, y_k$$
$$\text{s.t.} \quad \alpha y_j \leq \beta_0 + \sum_{i=1}^{2} \beta_i x_{ij} \, ; j=1,2,\ldots,N$$
$$\alpha, \beta_i \geq 0 \, ; \beta_0 \text{ free in sign}$$

Let asterisks denote optimal values and dual variables be transformed as $\hat{\beta}_0 = \beta_0 / \alpha^*$, where $\alpha^*$ is assumed to be positive. Then the dual problem reduces to the simpler form

$$\text{Max } \hat{\beta}_0$$
$$\text{s.t.} \quad y_j \leq \hat{\beta}_0 + \sum_{i=1}^{2} \hat{\beta}_i x_{ij}$$
$$\hat{\beta} \geq 0; \hat{\beta}_0 \text{ free in sign} \tag{3.1.4}$$

If the unit k is technically efficient then we have the production frontier

$$y_k = \hat{\beta}_0^* + \hat{\beta}_1^* x_{1k} + \hat{\beta}_2^* x_{2k} \tag{3.1.5}$$

To deal with the growth frontier we consider the rates of input and output growth as $\tilde{x}_{ij} = \Delta x_{ij} / x_{ij}$ and $\tilde{y}_j = \Delta y_j / y_j$ and set up the models corresponding to (3.1.1) through (3.1.3). For instance the input oriented model is

$$\text{Min } \tilde{\theta}$$
$$\text{s.t.} \quad \sum_{j=1}^{N} \tilde{x}_j \tilde{\lambda}_j \leq \hat{\theta} \tilde{x}_k; \Sigma \tilde{y}_j \tilde{\lambda}_j \geq \tilde{y}_k$$
$$\Sigma \tilde{\lambda}_j = 1, \hat{\lambda}_j \geq 0; j=1,2,\ldots,N \tag{3.1.6}$$

Its dual is

Max $\tilde{\beta}_0 - \tilde{\alpha}y_k$

s.t.    $\tilde{\alpha}y_j \leq \tilde{\beta}_0 + \sum\limits_{i=1}^{2} \tilde{\beta}_i \tilde{x}_{ij}$                    (3.1.7)

$\tilde{\alpha}, \tilde{\beta}_i \geq 0, \tilde{\beta}_0$ free in sign

If the unit k is growth efficient, then the growth frontier may be written as

$$\tilde{y}_k = \overline{\beta}_0^* + \overline{\beta}_1^* \tilde{x}_{1k} + \overline{\beta}_2^* \tilde{x}_{2k}$$                    (3.1.8)

where $\overline{\beta}_0 = \tilde{\beta}_0 / \tilde{\alpha}^*, \overline{\beta}_i = \tilde{\beta}_i / \tilde{\alpha}^*$ and it is assumed that $\tilde{\alpha}^*$ is positive at its optimal value. TFP growth along the growth frontier reduces to

$$\overline{\beta}_0^* = \tilde{y}_k - \overline{\beta}_1^* \tilde{x}_{1k} - \overline{\beta}_2^* \tilde{x}_{2k}$$                    (3.1.9)

If the production function exhibits constant returns to scale, then $\overline{\beta}_1^* + \overline{\beta}_2^* = 1$. Recently Hall (1990) developed a modified Solow residual by incorporating the presence of market power (i.e., degree of monopoly) and also overall increasing returns. But since the share of labor can be measured in relation to either total costs or total revenues, two concepts of modified Solow residuals are possible. Let $\mu$ be the mark-up ratio i.e., the ratio of output price to marginal costs and $\gamma$ be the overall returns to scale index, i.e., $\gamma_j = \sum\limits_{i=1}^{2} (\partial \ell n y_j / \partial \ell n x_{ij})$, then the two Solow residuals along the growth frontier for $DMU_k$ are as follows:

$$\tilde{\beta}_0^* = \tilde{y}_k - \sum\limits_{i=1}^{2} \tilde{\beta}_i^* \tilde{x}_{ik} - (\gamma_k - 1) \sum\limits_{i} \tilde{\beta}_i^* \tilde{x}_{ik}$$

and                                                                                      (3.1.10)

$$\tilde{\beta}_0^* = \tilde{y}_k - \sum\limits_{i=1}^{2} \tilde{\beta}_i^* \tilde{x}_{ik} - \tilde{\beta}_2^* (\mu_k - 1)(\tilde{x}_{1k} - \tilde{x}_{2k}) - (\gamma_k - 1)\tilde{x}_{2k}$$

Note that in both cases if perfect competition prevails (i.e. $\mu_k = 1$) and constant returns hold (i.e. $\gamma_k = 1$) then we have the traditional Solow residual as in (3.1.8). Hall's empirical estimates found significant evidence for the existence of market power ($\mu > 1$) and also increasing returns to scale ($\gamma > 1$) for several sectors of the U.S. economy.

In the DEA framework we may incorporate the effects of degree of monopoly and the overall returns to scale in two alternative ways. One is to set up the nonlinear model for obtaining the DEA estimates of $\tilde{\beta}_i$, $\mu_k$ and $\gamma_k$. For instance the DEA model (3.1.7) would appear as

$$\text{Max } \tilde{\beta}_0 - \tilde{\alpha} y_k$$

$$\text{s.t.} \quad \tilde{\alpha} y_j \leq \tilde{\beta}_0 + (\sum_i \tilde{\beta}_i^* \tilde{x}_{ij}) \gamma_j \qquad (3.1.11)$$

$$\tilde{\alpha}, \tilde{\beta}_i \geq 0, \tilde{\beta}_0 \text{ free in sign}$$

A second method would use a prior estimate of $\mu$ and $\gamma$ based on the industry-wide data. Thus if the degree of monopoly power is given for the industry as $\mu = \mu_k$ for all k, then one may use the second equation of (3.1.10) to set up the Solow residual in the DEA model as

$$\text{Max } \tilde{\beta}_0 - \tilde{\alpha} y_k$$

$$\text{s.t.} \quad \tilde{\alpha} y_j \leq \tilde{\beta}_0 + \sum_{i=1}^{2} \tilde{\beta}_i \tilde{x}_{ij} - \tilde{\beta}_2 (\mu - 1)(\tilde{x}_{1j} - \tilde{x}_{2j}) - (\gamma_j - 1) \tilde{x}_{2j}$$

$$\tilde{\alpha}, \tilde{\beta}_i, \gamma_j \geq 0, \tilde{\beta} \text{ free in sign.} \qquad (3.1.12)$$

In case the prior estimate of $\mu$ is not available one has to estimate the nonlinear term $(\tilde{\beta}_2(\mu - 1))$ from the DEA model above. But since $\tilde{\beta}_2$ is already computed in the term $\sum_i \tilde{\beta}_i \tilde{x}_{ij}$, the term $(\mu - 1)$ can be found out from the composite variable $b = \tilde{\beta}_2 (\mu - 1)$ determined from the programming model.

Recent trends in growth accounting literature has emphasized another aspect of technological change which is embodied in the physical capital input $(x_2)$. Thus Hulten (1992) and Greenwood et al (1997) have found for the U.S. economy over the period 1950-1994 that the investment-specific technological change explains close to 60% of the growth in output per hour worked. Residual, neutral productivity change (i.e., TFP growth) then accounts for the remaining 40%. The main feature of this growth process is that the production of capital goods becomes increasingly efficient with the passage of time. The introduction of the new and more efficient capital goods acts as an important source of productivity change which causes the relative prices of new equipments to decline. Concrete examples in support of this hypothesis abound: new and more powerful computers, faster and more efficient means of telecommunication and transportaiton and the flexible manufacturing systems in robotization in assembly lines and so on.

In order to show the impact of new vintage capital inputs, we decompose the input $x_2$ into two parts $x_3$ and $x_4$, where $x_4$ is new equipment and $x_3$ is structures. The growth of these two inputs follow the equations of motion as

$$x_3(t+1) = (1-\delta_3)x_3(t) + z_3(t), \quad 0 < \delta_3 < 1$$
$$x_4(t+1) = (1-\delta_4)x_4(t) + \tau z_4(t), \quad 0 < \delta_4 < 1 \tag{3.1.13}$$

where $\delta_3$, $\delta_4$ are fixed rates of depreciation, $z_3$, $z_4$ are investments and $\tau$ is the productivity of new equipment which grows over time through knowledge diffusion across firms and industries. Greenwood $et\ al$ have estimated the productivity term $\tau$ over U.S. data (1950-94).

One may incorporate these aspects of embodied technological change in an allocative DEA model, where the prices of $x_i$ and $z_i$ denoted by $q_i$ and $r_i$ are available through the market data or its forecasts. The efficient time paths of the various inputs and outputs may then be obtained by solving the following linear dynamic program

$$\text{Min } C(T) = \sum_{t=1}^{T} (1+\rho)^{-t}[q_1x_1 + q_3x_3 + q_4x_4 + r_3z_3 + r_4z_4]$$

$$\sum_{j=1}^{N} x_{ij}\lambda_j \le x_i; \quad i = 1,3,4$$

s.t. $\quad \sum_j z_{ij}\lambda_j \le z_i; \quad i = 3, 4 \tag{3.1.14}$

$$\sum_j y_j\lambda_j \ge y_k; \Sigma\lambda_j = 1, \lambda_j \ge 0$$

and the conditions (3.1.13) for $j = 1,2,...,N$

The model could be solved one period ahead with $T = 2$, or over a given horizon by means of the dynamic programming algorithm or by Pontryagin's maximum principle. The productivity parameter $\tau$ has to be estimated either extraneously for the whole industry or by a nonlinear programming algorithm.

Thus we may conclude this section by emphasizing the premise that the efficiency dynamics needs to be decomposed in terms of its various sources such as neutral technological change, market power, overall returns to scale and investment in new equipment, so that an extended DEA model may be set up for efficiency comparison among different units or firms.

## 3.2    Investment in Knowledge

Current research in the DEA field has paid very little attention to knowledge capital in the form of R&D inputs and cumulative experience associated with human capital. These inputs underlying knowledge capital have significant impacts on cost and productivity, that are quite different from other physical inputs. First of all, the human capital measured by cumulative experience and knowledge associated with new innovations and technology affects future productivity to a significant degree; furthermore, learning by doing generates substantial increasing returns to scale. The technology-intensive industries such as microelectronics, PC industry and semiconductors exemplify this productivity-enhancing effect of human capital. Secondly, these inputs have significant spillover effects on other firms in the industry; thus productivity in one firm entails its diffusion to other firms. This externality effect is inter-firm, though it develops initially as an intra-firm resource. Finally, viewing output of research as designs the human capital is nothing but the knowledge of how to work the blueprints. A single firm's research may facilitate and also be facilitated by the cumulative number of designs invented by all other firms in the industry. Thus the scale effects on the whole industry become important due to learning by doing.

In this section we extend the nonparametric efficiency approach of DEA models to incorporate the R&D and also analyze the dynamics of the learning by doing process.

Consider a standard input oriented DEA model for testing the relative efficiency of a reference firm or decision making unit h (DMU$_h$) in a cluster of N units, where each DMU$_j$ products s outputs ($y_{rj}$) with two types of inputs: m physical inputs ($x_{ij}$) and n R&D inputs as knowledge capital ($z_{wj}$):

$$\text{Min } \theta + \phi, \text{ subject to } \sum_{j=1}^{N} x_j\lambda_j \leq \theta X_h ; \sum_{j=1}^{N} Z_j\lambda_j \leq \phi Z_h$$

$$\sum_{j=1}^{N} Y_j\lambda_j \geq Y_h ; \sum_{j=1}^{N} \lambda_j \geq 0 ; j=1,2,\ldots,N \qquad (3.2.1)$$

Here $X_j$, $Z_j$ and $Y_j$ are the observed input and output vectors for each DMU$_j$, where $j=1,2,\ldots,N$. Let $\lambda^* = (\lambda_j^*)$, $\theta^*$, $\phi^*$ be the optimal solutions of model (3.2.1) with all slacks zero. Then the reference unit or firm h is said to be technically efficient if $\theta^* = 1.0 = \phi^*$. If however $\theta^*$ and $\phi^*$ are positive but less than unity, then it is not technically efficient at the 100% level, since it uses excess inputs measured by $(1-\theta^*)$ $x_{ih}$ and $(1-\phi^*)z_{wh}$. Overall efficiency (OE$_j$) of a unit j however combines the technical (TE$_j$) or production efficiency and the allocative (AE$_j$) or price efficiency as follows: OE$_j$ = TE$_j$ × AE$_j$. To measure overall efficiency of a DMU$_h$ one solves the cost minimizing model:

$$\text{Min } C = c' x + q' z$$
$$\text{s.t.} \quad X\lambda \leq x; Z\lambda \leq z; Y\lambda \geq Y_h; \lambda' e = 1; \lambda \geq 0 \qquad (3.2.2)$$

where e is a column vector with N elements each of which is unity, prime denotes transpose, c and q are unit cost vectors of the two types of inputs x and z which are now the decision variables and $X = (X_j)$, $Z = (Z_j)$ and $Y = (Y_j)$ are appropriate matrices of observed inputs and outputs. Denoting optimal values by asterisks, technical efficiency is now given by the $= \theta^* + \phi^*$. Overall efficiency ($OE_h$) is $C_h^* /(C_h (\theta^* + \phi^*))$.

Now consider the special characteristics of the research inputs. First of all, these R&D inputs lower the initial unit production cost $c_i$. Hence the total cost function in (3.2.2) may be written as

$$C = \sum_i \left[ (c_i - f_i)x_i + f_i^2 \right] + \sum_w q_w z_w \tag{3.2.3}$$

Here $f_i$ is the unit cost reduction with $f_i < c_i$ and the cost function $c(x_i, f_i) = (c_i - f_i)$ $x_i + f_i^2$ is assumed to be convex indicating diminishing returns of the underlying R&D production function. When $f_i$ is a decision variable, we may consider total R&D investment fund as a scalar variable and its allocation to each input as $f_i = \mu_i z_0$. In this case the constraints $Z\lambda \leq z$ are dropped along with the cost component $q'z$ and the allocation ratios $\mu_i$ are the new decision variables. This type of specification has been frequently adopted in the industrial organization literature. In this case the transformed DEA model becomes

$$\text{Min } C = \sum_i \left[ (c_i - \mu_i z_0)x_i + \mu_i^2 z_0^2 \right]$$

$$\text{s.t.} \quad X\lambda \leq x; Y\lambda \geq Y_h; \lambda'e = 1, \mu'e = 1 \tag{3.2.4}$$

$$\lambda \geq 0; \mu \geq 0$$

where $z_0$ can be set equal to unity without loss of generality. This is a quadratic programming model which can be solved for the optimal vectors $x^*, \lambda^*, \mu^*$ in order to test the efficiency of $DMU_h$. Clearly this model (3.2.4) would differ from the model (3.2.2) in two important respects. One is the learning effect due to a reduction in unit costs of production through research inputs. The second is the interdependence (i.e., complementarity) of the two types of inputs.

A second type of impact of research is through learning by doing in the sense of Arrow (1962), who considered cumulative gross investment as an index of experience or knowledge. Let $z(t) = (z_w(t))$ be the vector of investments and $\int_0^t z(t)dt$ be the cumulative gross investment where

$$dk(t)/dt = z(t) - \hat{d}k(t) \tag{3.2.5}$$

with $\hat{d} = (d_w)$ is a diagonal matrix with given depreciation rates $d_w$. In this case the transformed DEA model becomes dynamic as follows:

$$\text{Min} \int_0^\infty e^{-\rho t}[c'x(t) + C(z(t))]dt \qquad (3.2.6)$$

Here $C(z(t))$ is the adjustment cost of investment inputs which is geneally nonlinear in the current literature.

A third type of characterization of the research inputs and their productivity is in the current literature of new growth theory. Thus Lucas (1993) considered a growth process where each firm has a production function, where its output depends on its own labor and physical capital inputs as well as the total knowledge capital of the whole industry. The availability of industry's knowledge capital occurs through the spillover mechanism or diffusion of the underlying information process. The utilization of the industry's knowledge capital by each firm has been called by Jovanovic (1997) as the learning effect which is very significant in the modern software based industries. To characterize this learning effect we introduce a composite input vector $X_j^C$ for $DMU_j$ as the share of each $DMU_j$ out of the industry total supply of each input, e.g., $\sum_{j=1}^N X_{ij}^C = X_i^T$, where $X_i^T$ is the total industry supply of input i. We can then formalize the input-oriented DEA model in two forms as before:

$$\text{Min } \theta + \phi$$
$$\text{s.t.} \sum_{j=1}^N X_j\lambda_j \le \theta X_h; \sum_j X_j^C\lambda_j \le \phi X_h^C \qquad (3.2.7)$$
$$\sum_j Y_j\lambda_j \ge Y_h; \lambda'e = 1, \lambda \ge 0$$

and

$$\text{Min } C = c'x + q'x^C$$
$$\text{s.t. } X\lambda \le x; X^C\lambda \le x^C; Y\lambda \ge Y_h; \lambda'e = 1; \lambda \ge 0 \qquad (3.2.8)$$

On using the Lagrange multipliers $\alpha$, $\beta$, $\gamma$, and $\alpha_0$, the production frontier for DMUh in case of model (3.2.8) may be easily derived from the dual problem as

$$\alpha'Y_h = \beta'X_h + \gamma'X_h^C + \alpha_0; \alpha, \beta, \gamma \ge 0$$

Clearly the interdependence of the two inputs x and $x^C$ can be easily introduced in this framework through nonlinear interaction terms in the objective function (3.2.8) or, through the method used in (3.2.4) before.

Thus the generalized DEA models incorporate three additional sources of relative efficiency not found in the conventional DEA models: (1) unit cost reduction due to the complementarity effect of R&D inputs, (2) the increasing returns to scale due to learning by doing and finally (3) the spillover effect of knowledge capital in the industry as a whole.

An important area of learning by doing is the study of quality-based learning. This is the learning curve approach which suggests that effort devoted to improving the quality of manufactured products may reduce unit costs. Here experience is usually modeled as a cumulative product volume, where production is adjusted for quality. Thus the cumulative experience of output type r may be written as

$$\varepsilon_r(t) = \varepsilon_r(0) + \int_0^t y_r(s)Q_r(s)ds \qquad (3.2.9)$$

Unit costs of output r (r=1,2,...,s) are $c_r(\varepsilon_r, Q_r)$ depending on output volume and quality. If we assume that quality-based learning benefits accrue in durect production costs, then this can be written as

$$c_r(\varepsilon_r, Q_r) = c_r(\varepsilon_r) + c_r(Q_r)$$

Let $p_r = p_r(y_r)$ be the output price depending on output value $\varepsilon_r(t)$, so that the profit function is $\pi(t) = \sum_{r=1}^{s} [p_r - c_r(\varepsilon_r) - c_r(Q_r)]y_r(t)$. A transformed DEA model can now be set up which maximizes an adjusted profit function

$$\text{Max } J = \int_0^T e^{-pt}\pi(t)dt - \theta(t)$$

$$\text{s.t. } \sum_{j=1}^{N} y_{rj}\lambda_j(t) \geq y_r \quad (r=1,2,...,s)$$

$$\sum_{j=1}^{N} x_{ij}\lambda_j(t) \leq \theta(t)x_{ih} \qquad (3.2.10)$$

$$\sum \lambda_j(t) = 1, \lambda_j(t) \geq 0 \text{ and the constraint (3.2.9)}$$

to determine the optimal paths of output $(\varepsilon_r^*(t), y_r^*(t))$ and quality $Q_r^*(t)$ along with technical efficiency $\theta^*(t)$ and the optimal weights $\lambda_j^*(t)$. Clearly the maximization of J in (3.2.10) with respect to $\varepsilon_r(t)$ is a calculus of variations problem subject to inequality constraints but maximization of J with respect to $y_r(t)$ can be solved as a static quadratic programming model, provided the expected price variables $p_r$ are linear functions of output. When the prices $p_r$ are constants, the static problem reduces to a standard DEA model like (3.2.2), except that the discounted profit is maximized instead of the current input costs.

The Euler-lagrange condition for the optimal path of $\varepsilon_r(t)$ can be easily derived as

$$\rho \frac{\partial c_r(Q_r)}{\partial Q_r} + y_r(t) \frac{\partial c_r(\varepsilon_r)}{\partial \varepsilon_r} = \frac{\partial^2 c_r(Q_r)}{\partial Q_r^2} \dot{Q}_r \qquad (3.2.11)$$

where $\dot{Q}_r = dQ_r / dt$. Clearly the efficient level $y_r^*(t)$ of output depends now on the time rate of change of quality and the change in output volume. In the steady state one obtains

$$y_r^* = \rho MC(Q_r) / |MC(\varepsilon_r)|$$

where MC is marginal costs. Here $MC(\varepsilon_r) = \partial c_r(\varepsilon_r) / \partial \varepsilon_r$ is negative, since higher cumulative experience (output) reduces input costs. Thus the efficient output level $y_r^*$ in the steady state is higher, the higher the marginal cost reduction of output due to cumulative experience. Moreover one could easily derive from (3.2.11) the optimal quality path

$$\dot{Q}_r(t) = [\rho MC(Q_r) + y_r MC(\varepsilon_r)](\varepsilon_r) / \frac{\partial MC(Q_r)}{\partial Q_r}$$

which shows that $Q_r(t)$ declines over time whenever $MC(Q_r)$ increases or, $MC(\varepsilon_r)$ rises. Also if the discount rate $\rho$ is zero, $\dot{Q}_r(t)$ becomes negative, since $c_r(\varepsilon_r)$ is decreasing. On applying the transversality condition $e^{-rT} MC(Q_r(T))/y_r(T) = 0$, the convergence of $Q_r(t)$ to the steady state value $\bar{Q}_r$ may be directly estimated. Thus the dynamics of the optimal path of quality acquires an important meaning, i.e., the optimal quality level $Q_r(t)$ may start initially at a level higher than $\bar{Q}_r$ but eventually settles down to the steady state level $\bar{Q}_r$. This is due to the convexity of the quality cost function $c(Q_r)$.

Thus the quality-based learning when introduced into the dynamic DEA model allows us to analyze two additional sources of efficiency of a reference DMU or firm, e.g., (1) the optimal time path of quality changes and (2) the cumulative experience in the form of output volume. These dynamic aspects are ignored in the static DEA models, thus biasing their efficiency estimates.

## 3.3    Dynamics of Allocative Efficiency

The economist's view of Pareto efficiency is intimately connected with the competitive equilibrium, where market prices act as signals for the agents who are price takers. The concept of technical efficiency (TE) of a reference unit or $DMU_k$ ignores the price information, whereas the allocative efficiency (AE) allows optimal input-mix based on competitive market prices. Overall efficiency (OE) of a unit is therefore the product of TE and AE. Let the reference unit be $DMU_k$ and the observed input-output data be given by the input and output vectors $(X_j, Y_j)$ for $j=1,2,...,N$ with m inputs and s outputs respectively. In a static framework we set up the following LP model for testing the overall efficiency of the reference unit $DMU_k$:

$$\text{Min } q' x$$
$$\text{s.t. } \sum_{j=1}^{N} X_j \lambda_j \le x; \sum_{j=1}^{N} Y_j \lambda_j \ge Y_k \qquad (3.3.1)$$
$$\sum \lambda_j = 1, \lambda_j \ge 0; \; j=1,2,...,N$$

Now consider a dynamic extension of the overall efficiency model (3.3.1), where the reference unit $DMU_k$ or the firm k uses a quadratic loss function to choose the sequence of decision variables $x(t) = (x_i(t))$ over a planning horizon. The objective now is to minimize the expected present value of a quadratic loss function subject to the constraints of (3.3.1) as follows:

$$\underset{x(t),\lambda(t)}{\text{Min}} \; L = E_t \left\{ \sum_{t=1}^{\infty} \rho^t [q'(t)x(t) + (1/2)(d'(t)Wd(t)) + (1/2)(z'(t)Hz(t))] \right.$$
$$\text{s.t. } \sum_{j=1}^{N} X_j(t)\lambda_j(t) \le x(t); \sum_{j=1}^{N} Y_j(t)\lambda_j(t) \ge Y_k(t) \qquad (3.3.2)$$
$$\sum_{j=1}^{N} \lambda_j(t) = 1; x(t) \ge 1; \lambda(t) \ge 0$$

where $\rho$ is a known discount factor and the vectors $d(t) = x(t)—x(t-1)$ and $z(t) = x(t)—\hat{x}(t)$ are deviations with W and H being diagonal matrices representing the weights. The quadratic part of the objective function may be interpreted as adjustment costs, the first component being the cost of fluctuations in input usages and the second being a disequilibrium cost due to the deviations from the desired path denoted by $\hat{x}(t)$. On using the Lagrange multiplier $\mu(t) = (\mu_i(t))$ for the first constraint and assuming an interior optimal solutions with $x_i(t) > 0$, the optimal intertemporal path of inputs $x_i(t)$ may be specified as follows:

$$\alpha_i x_i^*(t) = w_i x_i^*(t-1) + \rho w_i x_i^*(t+1) + h_i \hat{x}_i(t) - q_i(t) + \mu_i^*(t)$$
$$i = 1,2,...,m \qquad (3.3.3)$$

where asterisk denotes optimal values, $\alpha_i = w_i + \rho w_i + h_i$ and it is assumed that future expectations are realized, i.e., $E_t(x_i(t+1)) = x_i(t+1)$. This last assumption is also called the rational expectations hypothesis, implying a perfect foresight condition. This type of hypothesis has been frequently used in recent macrodynamic models in economics. Several implications of this optimal linear decision rule (3.3.3) may now be briefly discussed.

First of all, if the observed input path $X_k(t)$ does not follow the optimal path $x^*(t)$ for any t, we have intertemporal inefficiency due to this divergence. This divergence may be cumulative, if it persists over several time points. Secondly, the myopic optimal vlaue $x^*$ computed from (3.3.1) can be directly compared with the intertemporal optimal path $x^*(t)$. Since the static efficiency associated with the myopic input levels $x^*$ ignores the potential losses arising over time, it is frequently biased. Hence the static efficiency ranking of N DMUs is most likely to be suboptimal in a dynamic setting. Thirdly, one could consider the linear decision rules (3.4.3) as a set of linear difference equations and taking the stable characteristic roots one could characterize the path of convergence to the steady state equilibrium values, e.g.,

$$\lim_{t \to \infty} x_i^*(t) = \overline{x}_i^*, \; I=1,2,\ldots,m \qquad (3.3.4)$$

Sengupta (1995) has discussed the implications of this convergence process in the theory of adjustment costs when technological change occurs in the expansion of capital inputs. Finally, the cost and production frontiers implicit in the static DEA model (3.3.1) may be updated at each time point t by incorporating the intertemporally optimal values. Since the linear decision rule (3.3.3) depends linearly on the parameters $w_i, \rho, h_i \hat{x}_i(t)$ and $q_i(t)$, one could explain the source of divergence of static from dynamic efficiency in terms of the changes in these parameter values. For example the higher the future levels of input price $q_i(t)$, the lower will be the optimal input demand $x_i^*(t)$. Thus any $DMU_k$ has to forecast the future values of the prices and costs $q_i(t)$, $w_i$, $h_i$ with a low forecasting error so that it may be on the dynamically efficient cost frontier. Otherwise a large error variance may cause large deviations from the expected intertemporal optimal path for $x(t)$. Thus when the agent is risk averse, he may consider the potential loss due to such large error variances and incorporate them in the original objective function of model (3.3.2). In this case the objective function gets transformed into more nonlinear than a quadratic.

Next we consider a second type of dynamic formulation, when capital inputs are distinguished from the current inputs. For simplicity we assume the first m-1 inputs to be current and the m-th input as capital. If $q_m(t)$ is the price of the capital input, then $q_m(t) x_m(t)$ is the investment in durable goods in the process. Assuming continuous discounting at an instantaneous rate r, the cost on current account of an initial investment outlay $q_m(t) x_m(t)$ is $r \, q_m x_m$. Hence the total current cost is

$$c = \sum_{i=1}^{m-1} q_i x_i + rq_m x_m \qquad (3.3.5)$$

Minimizing this cost funciton c in (3.3.5) subject to the constraints of model (3.3.1) provides a measure of overall efficiency in the short period. If x* is the optimal input vector, then the overall inefficiency of $DMU_k$ in the use of capital input is given by

$$OE_k(x_m) = \frac{rq_m x_m^*}{rq_m X_{mk}} = x_m^* / X_{mk}$$

When a planning horizon over the interval $0 \leq t \leq \infty$ is introduced, then the decision problem is one of choosing the current and capital inputs so as to minimize the total discounted cost

$$c = \int_0^\infty e^{-rt} \left[ \sum_{i=1}^{m-1} q_i(t)x_i(t) \right] dt + rq_m(t)x_m(t)$$

$$\text{s.t.} \sum_{j=1}^N x_{ij}\lambda_j(t) \leq x_i(t); \ i=1,2,\ldots,m-1$$

$$\sum_{j=1}^N x_{mj}\lambda_j(t) \leq x_m(t); \sum_{j=1}^N y_{rj}\lambda_j(t) \geq Y_{rk} \ ; r=1,2,\ldots,s \qquad (3.3.6)$$

$$\sum_{j=1}^N \lambda_j(t) = 1, x(t) \geq 0; \lambda(t) \geq 0$$

However this formulation (3.3.6) ignores the investment aspect of capital expansion. Let $z_m(t)$ be gross investment and $\delta$ the constant rate of depreciation for the expansion of capital inputs, i.e.,

$$\dot{x}_m(t) = z_m(t) - \delta x_m(t) \qquad (3.3.7)$$

where the dot denotes the time derivative. In this case the cost of investment $c(x_m(t))$ has to be included as a component of the objective function, where the last term has to be dropped. The final objective function thus becomes:

$$\text{Max } J = -c = -\int_0^\infty e^{-rt} \left[ \sum_{i=1}^{m-1} q_i(t)x_i(t) + c(z_m(t)) \right] dt$$
$$\text{s.t. (3.3.6) and (3.3.7)} \qquad (3.3.8)$$

This model helps to determine the optimal time path $z_m^*(t)$ of investment and hence the optimal time path of capital expansion in terms of $\{x_m^*(t), 0 < t < \infty\}$. This type of model can be easily solved by Pontryagin's maximum principle, whereby we introduce the Hamiltonian function H as:

$$H = e^{-rt}\left[\sum_{i=1}^{m-1} q_i(t)x_i(t) + c(z_m(t)) + p_m(z_m(t) - \delta x_m(t))\right]$$

where $p_m = p_m(t)$ is the adjoint function. If the optimal path of $x_m(t)$ exists, then by Pontryagin principle there must exist a continuous funciton $p_m(t)$ satisfying

$$\dot{p}_m(t) = (r + \delta)p_m(t) - \mu$$

where $\mu = \mu(t)$ is the Lagrange multiplier associated with the second constraint of model (3.3.8). Also we must have the optimal path of investment $z_m = (t)$ satisfying each moment of time t the optimality condition:

$$\partial c(z_m)/\partial z_m - p_m(t) \leq 0 \text{, at each t}$$

This implies for every positive level of investment the equivalence of marginal investment cost with the optimal shadow price, i.e.,

$$\partial c(z_m)/\partial z_m = p_m(t) \text{, for } z_m(t) > 0$$

In addition the adjoint variable $p_m(t)$ must satisfy the transversality condition

$$\lim_{t \to \infty} e^{-rt}p_m(t) = 0$$

i.e.    $$\lim_{t \to \infty} e^{-rt}p_m(t)x_m(t) = 0$$

If the investment cost function $c(z_m)$ is of a quadratic form, i.e., $c(z_m) = (1/2)\alpha z_m^2$, $\alpha > 0$, then the optimality conditions become

$$z_m^*(t) = p_m^*(t)/\alpha; \dot{x}_m^* = (p_m^*/\alpha) - \delta x_m^*$$
$$\dot{p}_m^* = (r + \delta)p_m^* - \mu^*; \lim_{t \to \infty} e^{-rt}p_m^*(t)x_m^*(t) = 0 \qquad (3.3.9)$$

with asterisks denoting optimal values. Note that the investment cost function includes potential losses from fluctuations in prices of capital and the cost of building capacity ahead of demand. The optimal trajectories $\{x_m^*(t), p_m^*(t); \; 0 < t < \infty\}$ determined by the system (3.3.9) of necessary conditions have several interesting economic implications for efficiency.

First of all, the steady state solutions $(\overline{x}_m^*, \overline{p}_m^*)$ on the optimal trajectory defined by (3.3.9) would be stable if the following conditions hold:

$$p_m^* / \alpha \underset{>}{\overset{<}{\gtrless}} \delta x_m^* \text{ according as } x_m \underset{<}{\overset{>}{\gtrless}} \overline{x}_m^*$$

and

$$(r + \delta)p_m \underset{>}{\overset{<}{\gtrless}} \mu^* \text{ according as } p_m \underset{<}{\overset{>}{\gtrless}} \overline{p}_m^*$$

since the differential equation above are linear. Also one could combine the two differential equations above to derive a single second order linear differential equation as follows:

$$\alpha \ddot{x}_m^* - \alpha r \dot{x}_m^* - \alpha \delta (r + \delta) x_m^* + \mu^* = 0 \tag{3.3.10}$$

Its characteristic equation is

$$u^2 - ru - \delta(r + \delta) = 0$$

which shows the two roots to be real and opposite in sign, i.e., $u_1 > 0$, $u_2 < 0$. Thus the steady state pair $(\overline{x}_m^*, \overline{p}_m^*)$ has the saddle point property. On assuming a fixed steady state value for $\mu^*$, the transient solution of equation (3.3.10) can be written as

$$x_m^*(t) = \left[ x_m^*(0) - \frac{\overline{\mu}^*}{\alpha\delta(r + \delta)} \right] e^{u_2 t} + \frac{\overline{\mu}^*}{\alpha\delta(r + \delta)}$$

where $x_m^*(0)$ is the initial value of $x_m^*(t)$ at $t = 0$. Note that the constant term $A_1$ in the solution

$$x_m^*(t) = A_1 e^{u_1 t} + A_2 e^{u_2 t} + \frac{\overline{\mu}^*}{\alpha\delta(r + \delta)}$$

has to be set equal to zero in order to satisfy the transversality conditions. Finally, if the observed path of capital expansion equals the optimal path, i.e., $x_m(t) = x_m^*(t)$ for every t, then the DEA model (3.3.6) would exhibit dynamic efficiency; otherwise any divergence of the two paths would generate inefficiency over time. In that case the conditional production function given an in optimal stock of capital inputs would exhibit myopic inefficiency.

## 3.4    Learning and Efficiency

Several types of learning in the productoin process have been discussed in the current literature. For example, Jovanovic (1997) has recently classified learning models into two broad types: one associated with technology and the other with human capital. We would consider here only the human capital aspect, which affects both quality and productivity improvements. Three types of measures of learning are used in our formulation. One is the cumulative research experience embodied in cumulative output, where the latter is very often taken as a measure of technological progress, e.g., the empirical studies of industrial productivity by Norsworthy and Jang (1992) have found the cost reducing effect of such technological progress to be substantial in microelectronics, telecommunications and similar other industries. The second measure is cumulative experience embodied in the strategic inputs such as capital goods in Arrow's model. The R&D expenditures allocated to improve the quality of any or all inputs may be considered here. Finally, the experience in 'knowledge capital' available to a firm due to a spillover from other firms may be embodied in the cost function of the firm through cumulative research inputs.

Consider now a standard input oriented DEA model for testing the relative efficiency of a reference firm or decision making unit h(DMU$_h$) in a cluster of N units, where each DMU$_j$ produces s outputs (y$_{rj}$) with two types of inputs: m physical inputs (x$_{ij}$) and n R&D inputs as knowledge capital (z$_{wj}$):

$$\text{Min } \theta + \phi, \text{ subject to } \sum_{j=1}^{N} X_j \lambda_j \le \theta X_h; \sum_{j=1}^{N} Z_j \lambda_j \le \phi Z_h$$

$$\sum_{j=1}^{N} Y_j \lambda_j \ge Y_h; \sum_{j=1}^{N} \lambda_j \ge 0 ; j=1,2,\dots,N \qquad (3.4.1)$$

Here X$_j$, Z$_j$ and Y$_j$ are the observed input and output vectors for each DMU$_j$, where j=1,2,...,N. Let $\lambda^* = (\lambda_j^*)$, $\theta^*$, $\phi^*$ be the optimal solutions of model (3.2.1) with all slacks zero. Then the reference unit or firm h is said to be technically efficient if $\theta^* = 1.0 = \phi^*$. If however $\theta^*$ and $\phi^*$ are positive but less than unity, then it is not technically efficient at the 100% level, since it uses excess inputs measured by $(1-\theta^*)$ x$_{ih}$ and $(1-\phi^*)$z$_{wh}$. Overall efficiency (OE$_j$) of a unit j however combines both technical (TE$_j$) or production efficiency and the allocative (AE$_j$) or price efficiency as follows: OE$_j$ = TE$_j$ × AE$_j$. To measure overall efficiency of a DMU$_h$ one solves the cost minimizing model:

$$\text{Min } C = c' x + q' z$$

$$\text{s.t.} \quad X\lambda \leq x; \; Z\lambda \leq z; \; Y\lambda \geq Y_h; \; \lambda' e = 1; \; \lambda \geq 0 \qquad (3.4.2)$$

where e is a column vector with N elements each of which is unity, prime denotes transpose, c and q are unit cost vectors of the two types of inputs x and z which are now the decision variables and $X = (X_j)$, $Z = (Z_j)$ and $Y = (Y_j)$ are appropriate matrices of observed inputs and outputs. Denoting optimal values by asterisks, technical efficiency is now given by the $= \theta^* + \phi^*$. Overall efficiency ($OE_h$) is $C_h^* / (C_h (\theta^* + \phi^*))$.

Now consider the special characteristics of the research inputs. First of all, these R&D inputs lower the initial unit production cost $c_i$. Hence the total cost function in (3.4.2) may be written as

$$\text{Min } TC = \sum_i \left[ (c_i - f_i (\sum_w q_w z_w)) x_i + \frac{1}{2} d_i x_i^2 \right] \qquad (3.4.3)$$
$$+ \frac{1}{2} \sum_{w=1}^{n} g_w z_w^2$$

subject to the constraints of model (3.4.2)

Here $f_i$ is the unit cost reduction with $f_i < c_i$ and the component cost functions are assumed to be strictly convex implying diminishing return to the underlying R&D production function. The optimal solutions $z_w$, $x_i$ and $\lambda_i$ now must satisfy the Kuhn-Tucker necessary conditions as follows:

$$f_i q_w x_i + \gamma_w \leq g_w z_w \, ; z_w \geq 0$$
$$f_i (\Sigma q_w z_w) + \beta_i \leq c_i + d_i x_i \, ; x_i \geq 0 \qquad (3.4.4)$$

If the unit ($DMU_h$) is efficient with positive input levels and zero slacks, then we must have equality $\partial L / \partial z_w = 0 = \partial L / \partial x_i$ where L is the Lagrangean function. Hence we can write the optimal values $(z_w^*, x_i^*)$ as:

$$z_w^* = (f_i q_w x_i^* + \gamma_w^*) / g_w \, ; w = 1.2, ..., n$$
$$x_i^* = (g_w z_w^* - \gamma_w^*) / (f_i q_w) \, ; i = 1, 2, ..., m \qquad (3.4.5)$$

By duality the production frontier for unit j=1,2,...,N satisfies

$$\alpha^{*'} Y_j \leq \alpha_0^* + \beta^{*'} X_j + \gamma^{*'} Z_j \, ; (\alpha^*, \beta^*, \gamma^*) \geq 0$$

where the equality holds if unit j is efficient and there is no degeneracy due to congestion costs. Clearly a negative (positive or zero) value of $\alpha_0^*$ implies increasing (diminishing constant) returns to scale.

Note that this generalized quadratic programming model (3.4.3) has many flexible features comapred to the traditional Farrell-type DEA model (3.4.2). First of all, if the research inputs are viewed as cumulative stream of past investment as in Arrow model of learning by doing, then the cost funciton TC in (3.4.3) may be viewed as a long run cost function. Given the capital input z* the reference firm solves for the optimal current inputs $x_i^*$ through minimizing the short run cost function TC(x|z*). Second, the learning effect parameter $f_i > 0$ shows that the efficiency estimates through DEA model (3.4.2) would be biased if it ignores the learning parameters. Third, the complementarity (i.e., interdependence) of the two types of inputs is clearly brought out in the linear relation between $x_i^*$ and $z_i$ in (3.4.5). For example, it shows that

$$\partial x_i^* / \partial z_w^* > 0, \partial x_i^* / \partial f_i > 0, \partial x_i^* / \partial \beta_i^* > 0$$

and

$$\partial z_w^* / \partial x_i^* > 0, \partial z_w^* / \partial f_i > 0, \partial z_w^* / \partial \gamma_w^* > 0$$

Finally, compared to a linear program this quadratic programming model (3.4.3) permits more substitution among the inuts, thus making it possible for more units to be efficient.

One limitation of the long run cost (3.4.3) minimization model above is that it ignores the time profile of output generated by cumulative investment experience. Let $z(t) = (z_w(t))$ be the vector of gross investments and $k(t) = \int_0^t z(s)ds$ be the cumulative value where

$$\dot{k}_w(t) = z_w(t) - \delta_w k_w(t) \qquad (3.4.6)$$
$$\delta_w : \text{fixed rate of depreciation}$$

In this case the transformed DEA model becomes dynamic as follows:

$$\text{Min} \int_0^\infty e^{-\rho t}[c'(t)x(t) + C(z(t))]dt$$

$$\text{subject to (3.4.6) and the constraints of model (3.4.2)}$$

Here $C(z(t))$ is a scalar adjustment cost, which is generally assumed nonlinear in the theory of investment. This type of formulation in the DEA framework recently analyzed by Sengupta (1995) shows the stability and adaptivity aspects of convergence to the optimal path.

Another type of characterization of the research inputs and their productivity is in the current literature of new growth theory. Thus Lucas (1993) considered a growth process where each firm has a production function, where its output depends on its own labor and physical capital inputs as well as the total knowledge capital of the whole industry. The availability of industry's knowledge capital occurs through the spillover mechanism or diffusion of the underlying information process. The utilization of the industry's knowledge capital by each firm has been called by Jovanovic (1997) as the learning effect which is very significant in the modern software based industries. To characterize this learning effect we introduce a composite input vector $X_j^C$ for $DMU_j$ as the share of each $DMU_j$ out of the industry total supply of each input, e.g., $\sum_{j=1}^{N} X_{ij}^C = X_i^T$, where $X_i^T$ is the total industry supply of input i. We can then formalize the input-oriented DEA model in two forms as before:

$$\text{Min } \theta + \phi$$
$$\text{s.t. } \sum_{j=1}^{N} X_j \lambda_j \leq \theta X_h ; \sum_j X_j^C \lambda_j \leq \phi X_h^C \qquad (3.4.7)$$
$$\sum_j Y_j \lambda_j \geq Y_h ; \lambda' e = 1, \lambda \geq 0$$

and

$$\text{Min } C = c'x + q'x^C$$
$$\text{s.t. } X\lambda \leq x ; X^C \lambda \leq x^C ; Y\lambda \geq Y_h ; \lambda' e = 1; \lambda \geq 0 \qquad (3.4.8)$$

On using the Lagrange multipliers $\alpha, \beta, \gamma$ and $\alpha_0$, the production frontier for $DMU_h$ in case of model (3.4.8) may be easily derived from the dual problem as

$$\alpha' Y_h = \beta' X_h + \gamma' X_h^C + \alpha_0 ; \alpha, \beta, \lambda \geq 0$$

Clearly the interdependence of the two inputs x and $x^C$ can be easily introduced in this framework through nonlinear interaction terms in the objective function (3.4.8) or, through the method used in (3.4.3) before.

Thus the generalized DEA models incorporate three additional sources of relative efficiency not found in the conventional DEA models: (1) unit cost reduction due to the complementarity effect of R&D inputs, (2) the increasing returns to scale due to learning by doing and finally (3) the spillover effect of knowledge capital in the industry as a whole. An important area of learning by doing is the study of quality-based learning. This is the learning curve approach which suggests that effort devoted to improving the quality of manufactured products may reduce unit costs. Here experience is usually modeled as a cumulative product volume, where production is adjusted for quality. Thus the cumulative experience of output type r may be written as:

$$\varepsilon_r(t) = \varepsilon_r(0) + \int_0^t y_r(s)Q_r(s)ds \qquad (3.4.9)$$

Unit costs of ourput r (r=1,2,…,s) are $c_r(\varepsilon_r, Q_r)$ depending on output volume and quality. If we assume that quality-based learning benefits accrue in direct production costs, then this can be approximated in a separable form as

$$c_r(\varepsilon_r, Q_r) = c_r(\varepsilon_r) + c_r(Q_r)$$

Let C(t) be the total cost function $C(t) = \sum_{r=1}^{s} [c_r(\varepsilon_r) + c_r(Q_r)]y_r(t)$. A transformed DEA model can now be set up which minimizes the discounted cost function with a finite horizon [0,T],

$$\text{Min } J = \int_0^T e^{-\rho t}C(t)dt + \theta(t)$$

$$\text{s.t. } \sum_{j=1}^{N} y_{rj}\lambda_j(t) \geq y_r \ (r=1,2,…,s)$$

$$\sum_{j=1}^{N} x_{ij}\lambda_j(t) \leq \theta(t)x_{ih} \qquad (3.4.10)$$

$$\sum \lambda_j(t) = 1, \lambda_j(t) \geq 0 \text{ and the constraint (3.4.9)}$$

to determine the optimal paths of output $(\varepsilon_r^*(t), y_r^*(t))$ and quality $Q_r^*(t)$ along with technical efficiency score $\theta^* = \theta^*(t)$ and the optimal weights $\lambda_j^*(t)$. Clearly the minimization problem in (3.4.10) is a calculus of variations problem subject to inequality constraints. Let L be the Lagrangean function

$$L = e^{-\rho t}\left[ -\sum_{r=1}^{s} y_r(c_r(\varepsilon_r) - c_r(Q_r)) - \theta + \beta'(\theta X_h - \sum_{j=1}^{N} X_j\lambda_j) \right.$$

$$\left. + \alpha'\left(\sum_{j=1}^{N} Y_j\lambda_j - y\right) + \alpha_0(1 - \lambda'e) + \sum_{r=1}^{s} \mu_r(y_rQ_r - \dot{\varepsilon}_r) \right] \qquad (3.4.11)$$

then the necessary conditions of optimality can be written in two parts: a dynamic part and a static part, e.g.,

$$\dot{\mu}_r = \rho\mu_r - MC_r(\varepsilon_r)y_r$$
$$\mu_r(T)e^{-\rho T} = 0; r = 1, 2, ..., s \qquad (3.4.12)$$

and

$$\mu_r Q_r \le c_r(Q_r) + c_r(\varepsilon_r) + \alpha_r$$
$$\alpha' Y_j \le \beta' X_j + \alpha_0; \beta' X_j = 1 \qquad (3.4.13)$$
$$\alpha, \beta \ge 0; \alpha_0 \text{ free in sign}; r=1,2,...,s$$

Since $Q_r = \dot{\varepsilon}_r / y_r$ this quality-based learning model must also satisfy the following dynamic equation

$$\frac{\partial MC_r(Q_r)}{\partial Q_r}\dot{Q}_r = \rho MC_r(Q_r) + MC_r(\varepsilon_r)y_r \qquad (3.4.14)$$

Here $MC_r(Q_r)$ and $MC_r(\varepsilon_r)$ are the marginal costs of quality improvement and cumulative experience respectively. Note that the experience cost function $C_r(\varepsilon_r)$ is assumed here to be positive, decreasing and convex, i.e., the experience gained from larger volume of output helps the firm reduce its production cost. Hence $MC_r(\varepsilon_r)$ is negative. The quality cost function $c_r(Q_r)$ is asumed to be positive and strictly convex. Thus $MC_r(Q_r)$ is increasing in the relevant domain. In order to get specific results assume the cost functions as follows:

$$MC_r(Q_r) = n_r Q_r; MC_r(\varepsilon_r) = -m_r, m_r > 0$$

where $n_r$ and $m_r$ are positive constants. Then the dynamic path of optimal quality improvement (3.4.14) becomes

$$\dot{Q}_r(t) = \rho Q_r(t) - y_r(t)(m_r / n_r) \qquad (3.4.15)$$

Let T tend to infinity. Then the steady state solution of optimal quality $Q_r^*$ reduces to

$$Q_r^* = m_r / y_r^* /(\rho n_r) \qquad (3.4.16)$$

This shows that in the steady state optimal qulaity $(Q_r^*)$ and optimal output $(y_r^*)$ are positively correlated and $\partial Q_r^* / \partial n_r < 0, \partial Q_r^* / \partial \rho < 0, \partial Q_r^* / \partial m_r > 0$. Furthermore, the steady state quality level $Q_r^*$ is stable provided the following conditions hold

$$y_r > \rho n_r Q_r / m_r \text{ if } Q_r > Q_r^*$$

and                                                                                              (3.4.17)

$$y_r < \rho n_r Q_r / m_r \text{ if } Q_r < Q_r^*$$

It is clear that under these conditions (3.4.17), whatever the initial value Q(0) of the quality level, the optimal system would convege to the steady state $Q_r^*$ in (3.4.16).

Define the quality-adjusted output $y_r Q_r$ by $\tilde{y}_r$. Then by differentiating $\dot{Q}_r(t)$ in (3.4.15) with respect to time one obtains after some algebraic manipulation the following optimal relation along the optimal quality path

$$\dot{\tilde{y}}_r(t) = (1/\rho)\left[ y_r \ddot{Q}_r(t) + \dot{y}_r Q_r(t)\left\{\frac{\rho}{y_r} + \frac{m_r}{n_r}\right\}\right]$$

which yields

$$\dot{\tilde{y}}_r(t)/\tilde{y}_r(t) = (1/\rho)\left[\frac{\ddot{Q}_r}{Q_r} + \frac{\dot{y}_r(t)}{y_r(t)}\left\{\frac{\rho}{y_r(t)} + \frac{m_r}{n_r}\right\}\right]$$                    (3.4.18)

This is an important result along the quality frontier. First of all, it shows that the growth rate of quality adjusted output rises when $\ddot{Q}_r$ and $\dot{y}_r$ increase. Even when $\ddot{Q}_r(t)$ is negative, this negative impact may be offset by a positive growth rate of output $g_r = \dot{y}_r(t)/y_r(t) > 0$. Secondly, the higher the discount rate, the lower the growth rate $\tilde{g}_r = \dot{\tilde{y}}_r(t)/\tilde{y}_r(t)$ of quality-adjusted output; a similar effect holds for higher (lower) values of $n_r$ ($m_r$). Finally, let $q_r(t)$ denote the growth rate of quality, i.e., $q_r(t) = \dot{Q}_r(t)/Q_r(t)$ then it follows that $\tilde{g}_r(t)$ is strongly positively correlated with $q_r(t)$, since

$$\tilde{g}_r(t) = (1/\rho)\left[\dot{q}_r(t) + q_r^2(t) + g_r(t)\left\{\frac{\rho}{y_r(t)} + \frac{m_r}{n_r}\right\}\right]$$

i.e., quality growth affects output growth and vice versa.

Note that if we drop the quality factor by setting $Q_r(t) = 1.0$ we obtain from (3.4.12) the necessary conditions along the optimal frontier, e.g.,

$$\dot{\mu}_r = \rho\mu_r - MC_r(\varepsilon_r)y_r; \lim_{T\to\infty} e^{-\rho T}\mu_r(T) = 0$$

$$\mu_r(t) \le c_r(\varepsilon_r) + \alpha_r \tag{3.4.19}$$

$$\alpha'Y_j \le \beta'X_j + \alpha_0; \beta'X_h = 1$$

By the convexity conditions on the cost function, these conditoins are also sufficient. In the steady state one obtains

$$y_r^* = \rho\mu_r^* / MC_r(\varepsilon_r^*)$$

showing that the otimal output level $(y_r^*)$ increases (decreases) as the marginal cost $MC_r(\varepsilon_r^*)$ of cumulative experience declines (increases). Note that if the firm h is technically efficient we must have

$$\alpha^{*\prime} Y_h = \beta^{*\prime} X_h = \alpha_0^*; \beta^{*\prime} X_h = 1 \tag{3.4.20}$$

where asterisks denote optimal values. But for dynamic efficiency it must also satisfy

$$\mu_r^*(t) = c_r(\varepsilon_r^*) + \alpha_r^*$$

$$\dot{\mu}_r^*(t) = \rho\mu_r^*(t) + MC_r(\varepsilon_r^*)y_r^*(t) \tag{3.4.21}$$

$$\lim_{T\to\infty} e^{-\rho T}\mu_r^*(T) = 0$$

Thus it is clear that the optimal weight $\alpha_r^*$ in $\alpha^{*\prime} Y_h$ determined by the static production frontier (3.4.20) is not fully optimal unless it satisfies the first equation (3.4.21) where $\alpha_r^*$ depends on the dynamic shadow price $\mu_r^*(t)$ of output. But $\mu_r^*(t)$ must be on the dynamic optimal path specified by the second equation of (3.4.21) which is often called the perfect foresight condition. Thus static efficiency affords only a partial view of total efficiency, where the production frontier of $DMU_h$ is to be written as

$$\sum_{r=1}^{s} \alpha_r^*(\mu_r^*)y_{rh} = \beta^{*\prime} X_h + \alpha_0^*; \beta^{*\prime} X_h = 1$$

where $\alpha_r^*(\mu_r^*)$ depends on $\mu_r^*$ which is determined at any time t by the perfect foresight condition.

Thus the quality-based learning when introduced into the dynamic DEA model through reductions in unit costs of production allows us to analyze seveal additional sources of efficiency of a reference firm or DMU, e.g., (1) the optimal path of quality changes, i.e., the convergence characteristics of the quality frontier, (2) the role of cumulative expeience in the form of output volume, and (3) the influence of the discount rate and the perfect foresight condition.

Now we consider learning by doing in the growth model of Lucas (1993), where current output depends on current inuts and the cumulative output in a nonlinear fashion. For instance with one input (x) and one output (y), the production function takes the form

$$y(t) = A y_c^{\delta}(t) x^{\beta} \tag{3.4.22}$$

where $y_c = \int_0^t y(s) ds$ is cumulative output, A is a positive constant, $\delta (0<\delta<1)$ is a learning parameter and $\beta$ is a scale parameter, where $0<\beta<1$ implies that for a given volume of output, discounted costs decline with the time horizon. Taking logs of both sides and then time differentiation yields the result

$$\dot{y}(t)/y(t) = (\dot{A}/A) + \delta(\dot{y}_c/y_c) + \beta(\dot{x}/x) \tag{3.4.23}$$

This shows directly how the growth of cumulative output $(\dot{y}_c/y_c)$ and the growth of current input $(\dot{x}/x)$ affect the growth of current output $(\dot{y}(t)/y(t))$. This type of use of learning curves and output rates has been extensively used in managerial efficiency studies, e.g., Womer (1979), Spence (1981) and Fine (1986). Recently Norsworthy and Jang (1992) used this approach to measure the productivity of technological change.

Consider now an extension of the loglinear model (3.4.23) to multiple inputs and multiple outputs in a DEA framework. Define for $DMU_j$ the following variables where $y_{rj}$ is cumulative output and $\dot{y}_{rj}$ is current output:

$$g_{rj} = \dot{y}_{rj}/(y_{rj})^{\gamma_r}, \tilde{g}_{rj} = \ln g_{rj}, \tilde{x}_{ij} = \ln x_{ij}$$

for N DMUs (j=1,2,...,N) and assume $\dot{y}_{rj}$ to be positive (which can always be done by adding a large positive constant). Then set up the loglinear DEA model for choosing optimal inputs $\tilde{x}_i$ with given input cost $c_i$

$$\text{Min} \sum_{i=1}^{m} c_i \tilde{x}_i$$

$$\text{s.t.} \sum_{j=1}^{N} \tilde{g}_{rj} \lambda_j \geq \tilde{g}_{rh} \quad (r=1,2,\ldots,s) \tag{3.4.24}$$

$$\sum_{j=1}^{N} \tilde{x}_{ij} \lambda_j \leq \tilde{x}_i \quad (i=1,2,\ldots,m)$$

$$\sum \lambda_j = 1, \lambda_j \geq 0$$

On using the dual variables $\alpha_r$, $\beta_i$ and $\alpha_0$ one obtains the optimality conditions

$$\sum_{r=1}^{s} \alpha_r (\ln \dot{y}_{rj} - \gamma_r \ln y_{rj}) \leq \sum_{i=1}^{m} \beta_i \ln x_{ij} - \alpha_0 \tag{3.4.25}$$

$$\alpha = (\alpha_r) \geq 0; \beta = (\beta_i) \geq 0; \gamma = (\gamma_r) \geq 0; \ \alpha_0 \text{ free in sign}$$
$$\beta_i < c_i \quad (i=1,2,\ldots,m)$$

Note that if the learning parameter $\gamma_r$ is either known or equal to unity for each r, then this problem (3.4.24) is a linear programming problem, otherwise a nonlinear problem has to be solved for obtaining the optimal values $\alpha_r^*, \beta_i^*$ and $\gamma_r^*$. Clearly if DMU$_h$ is overall efficient as in model (3.4.2), then we must have

$$\sum_r \alpha_r^* (\ln \dot{y}_{rh} - \gamma_r^* \ln y_{rh}) = \sum_i \beta_i^* \ln x_{ih} + \alpha_0^* \tag{3.4.26}$$

With one output this reduces to

$$\dot{y}_{1h} = A y_1^{\gamma_1^*} \prod_{i=1}^{m} x_{ih}^{\beta_i^*/\alpha_1^*}; \ln A = \alpha_0^* / \alpha_1^*$$
$$\alpha_1^* \ln \dot{y}_{rh} = \alpha_1^* \gamma_1^* \ln y_{rh} + \sum_i \beta_i^* \ln x_{ih} + \alpha_0^* \tag{3.4.27}$$

Note that if there were no learning parameter then the overall efficiency for DMU$_h$ would require

$$\alpha_1^* \ln \dot{y}_{rh} = \sum_i \beta_i^* \ln x_{ih} + \alpha_0^* \tag{3.4.28}$$

Thus the term $\alpha_1^* \gamma_1^* \ln y_{rh}$ expresses the efficiency gap due to learning. This gap may sometimes be very substantial, since $y_{rh}$ is cumulative output and the experience factor in some technology-intensive industries has a very significant impact.

Clearly technical efficiency defined before in the LP model (3.4.1) may be easily incorporated here as follows:

Min θ

$$\text{s.t.} \sum_{j=1}^{N} \tilde{g}_j \lambda_j \geq \tilde{g}_h ; \sum_{j=1}^{N} \tilde{X}_j \lambda_j \leq \theta \tilde{x}; \lambda'e = 1; \lambda \geq 0$$

It is clear that technical efficiency would depend here on the learning parameters also. Thus learning and growth efficiency are closely connected in the growth dynamics of efficient firms or DMUs.

## 3.5    Efficiency Persistence

Farrell (1957) who first developed the LP method of determining the efficiency frontier as the conical hull of the production surface mentioned some dynamic issues related to the capital inputs which yield outputs for several future periods. Two concepts he emphasized are particularly relevant here. One is the concept of an efficiency distribution over different units or firms, where efficiency is defined as the ratio of the observed to the optimal or maximal output. The empirical shape of this efficiency distribution and its change over time characterize in his view the dynamic elements of total factor productivity. The second concept used by Farrell is termed 'structural efficiency', which measures the extent to which an industry keeps up with the performance of its own best practice firms. Thus it provides a measure at the industry level of the extent to which firms are of optimum size, i.e., to which total industry productoin is optimally allocated between the constituent firms in the short run. Farrell thought that this concept has the greatest practical importance in comparing the efficiency of two or more industries or, even two or more economic systems.

Our objective in this section is two-fold: to develop a method of comparing dynamic efficiency over time in Farrell's framework and to provide a measure of persistence in productivity dynamics. The first captures the short run aspects of productivity change, while the second emphasizes the long run productivity shifts.

Consider the production frontier for a single firm producing one output (y) with m inputs $(x_i)$

$$y^* = f(x_1, x_2, \ldots, x_m) \tag{3.5.1}$$

where $y^*$ is he maximum possible output the firm can produce by using the m inputs in a technically efficient way. Consequently any observed output (y) is either on or below the production frontier, i.e., $y_j \leq y_j^*$, where $j=1,2,\ldots,N$ denotes any one of the N firms in the industry. Given the observed data on inputs $(x_{ij})$ and output ($y_j$), one may adopt two diferent methods for estimating the production frontier. One is parametric and the other nonparametric. In the parametric approach a specific form of the production frontier (3.5.1) is assumed,e.g., a loglinear form and then the parameters estimated by statistical methods. For example assuming linearity the model may be written as

$$y_j = \sum_{i=0}^{m} \beta_i x_{ij} - u_j, u_j \geq 0, j = 1, 2, \ldots, N \tag{3.5.2}$$

where $y_j^* = \sum_{i=0}^{m} \beta_i x_{ij}$ denotes the optimal output with $x_{0j}$ set equal to one, so that $\beta_0$ is the intercept term. The econometric approach to the production frontier estimates the parameter vector $\beta = (\beta_i)$ by the generalized method of moments or by the corrected least squares method, provided the statistical distribution of the error term is known. See, e.g., the survey by Sengupta (1995a, 1996a-e). The Farrell model adopts the nonparametric approach, where the statistical distribution of the error term u is not assumed. It evaluates the production frontier by a sequence of linear programming (LP) models so as to test if each firm is technically efficient or not. This is done by solving for each reference unit or firm k the following linear programmnig model:

$$\operatorname*{Min}_{\beta} g_k = \sum_{i=0}^{m} \beta_i x_{ik}$$
$$\text{s.t.} \ \sum_{i=0}^{m} \beta_i x_{ij} \geq y_j, j = 1, 2, \ldots, N \tag{3.5.3}$$
$$\beta_i \geq 0, i = 0, 1, \ldots, m$$

where the inputs of the reference unit k is used in the objective function. On transforming the parameters $\beta_i$ to $\gamma_i = \beta_i y_k$, where $y_k$ is the output level of the reference unit, the LP model (3.5.3) can be written in vector matrix terms as:

$$\operatorname*{Min}_{\gamma} g_k = \gamma' A_k$$
$$\text{s.t.} \ \gamma' A \geq y_k e', \gamma \geq 0 \tag{3.5.4}$$

where prime denotes transpose, $A = (a_{ij})$ is the input coefficient matrix with $a_{ij} = x_{ij}/y_j$ and $A_k = (a_{ik})$ as the column vector and e is a column vector with N elements having each element unity. Let $\gamma^*(k)$ be the optimal solution vector of the LP model (3.5.4) above. Then the reference unit or firm k is technically efficient if it holds that

$$y_k^* = \sum_{i=0}^{m} \gamma_i^* a_{ik} = y_k \tag{3.5.5}$$

and $s_k^* = y_k^* - y_k = 0$, where $s_k^*$ is the optimal slack variable represnting the excess of potential or optimal output $y_k^* = \gamma^{*\prime} A$ over actual output. By varying k in the index set $I_N = \{1,2,...,N\}$, one can geneate the whole efficiency surface. This allows a partition of the set $I_N$ into two subsets, one containing the units that are efficient by the criterion (3.5.5) and the other containing inefficient units. Let $E_1$ and $E_2$ be these two subsets of the observation set $I_N$, where $E_1$ contains the efficient units only. Based on these two subsets Farrell cmoputed the empirical frequency distribution of efficiency (e) defined by the ratio $e_j = y_j / y_j^*$, where $0 < e_j \leq 1.0$ for each j=1,2,...,N. Let F(e,t) be the cumulative frequency distribution of efficiency for a fixed time point t. An important issue in the analysis of productivity behavior is whether firms or units occupy a fixed rank in terms of their productivity aspects, so that the notion of a "representative firm" is useful, or their rank is subject to continuous variation. One promising way of looking at this problem is through the device of productivity transitions.

For a given t, one can arrange the efficiency ratios ($e_j$) in deciles starting from the lowest to the highest, i.e., rank one denotes the lowest efficiency level (lowest value of $e_j$) and rank 10 the highest. A transition probability table can then be constructed from the values of $\{e_j\}$ by assigning a firm to a decile in the cross-sectional distribution of efficiency in each year, based on the estimated value of its efficiency measure as determined by the LP model in DEA framework. One can also tabulate the incidence of transition of firms from a given decile in year t to the next decile (up or down) in year t+1. Let $d_t$ be a ten-element column vector representing the probability of occupying the ten decile position at time t, i.e., it contains the proportion of units in the ten decile classifications of the efficiency ratio $e_j$ at time t. The $\tau$-period transition can then be written as

$$d_{t+\tau} = A_\tau d_t \qquad\qquad (3.5.6)$$

Note that the (i,j) element of the transition matrix $A_\tau$ is the proportion of units or firms making the transition from decile i to decile j over $\tau$ periods for i,j = 1,2,...,10. The linear equations system (3.5.6) becomes a relatively simple Markovian scheme if we have that $A_\tau = A_1^\tau$, where $A_1$ is the one period transition matrix and it is estimated from the experience of firms over the entire period T (e.g., T = 10 in our empirical application). For simplicity we assume a Markovian scheme with one period transition matrix $A_1$ which we not denote by A.

Three important features of the transition probability model: $d_{t+1} = A d_t$ derived from Farrell measures of efficiency may now be commented upon. First of all, this generalizes the Farrell model, since the Farrell model can be used in every year t to identify units which are efficient and then in the second stage one can test over time if the decile containing the efficient units has retained its efficiency. For example let $d_{i,t}$ denote the proportion of firms or units in the highest efficiency level (i = rank 10), whch includes all units having estimated efficiency greater than or equal to 0.95, say $(0.95 \leq \hat{e} \leq 1)$. Since the data have errors of observation we include the interval $0.95 \leq \hat{e} \leq 1$ in our classification. One could then set up a linear regression model

$$d_{i,t+1} = \lambda_i d_{i,t} + a + \varepsilon_{i,t+1}, 0 \leq \lambda_i \leq 1 \qquad (3.5.7)$$

to estimate the parameter $\lambda_i$ representing persistence. The closer $\lambda_i$ is to one, the stronger the persistence. The error terms $\varepsilon_{i,t+1}$ can also be used to test for the existence of autocorrelation of the residuals. In general the regression system (3.5.7) can be written as

$$d_{i,t+1} = \sum_{j=1}^{10} \lambda_{ji} d_{i,t} + \varepsilon_{i,t+1} \qquad (3.5.8)$$

with

$$\sum_{i=1}^{10} \lambda_{ji} = 1, \lambda_{ji} \geq 0 , I,j=1,2,\ldots,10$$

Note that the off-diaponal elements of the transition matrix $\lambda_{ji}$ specify the probability of switching to another level of efficiency, higher or lower. Hence if it holds for the highest productivity rank i =10 that

$$\lambda_{ii} > \sum_{\substack{j \neq i \\ j=1}}^{10} \lambda_{ij}$$

then this provides strong support to the efficiency persistence hypothesis.

A second important feature of the transition dynamics approach in Farrell efficiency is that the efficiency ratio $e_j$ can be estimated from cumulative outputs rather than current outputs. Thus on a moving two-year interval basis, the inputs and outputs may be redefined in the LP model. Recently, Dalen (1993) has applied this method to panel data of 123 firms over four years in order to test for technical change in DEA framework. Since dynamic adjustments due to technological change may not be all completed in one year, this two-year cumulative values allow more adaptivity to the firms. Finally, the relationship between age and productivity can be analyzed in terms of the transition dynamics viewed not in terms of deciles but in terms of age groups and size groups. Thus one could test if the older firms (or large firms) retain their productive efficiency over time. This has implicatoins for long run industry efficiency.

The concept of 'structural efficiency' is another important measure proposed by Farrell, which can be generalized in dynamic terms. This efficiency measures the degree to which an industry keeps up with the performance of its own best practice firms. Let $N_1$ be the number of best practice firms or units in the industry comprising N firms in year t. Then the proportion $p_1(t) = N_1(t) / N(t)$ can be used as a measure of the degree to which the given industry keeps up with the performance of its best practice firms. Since the input output data contain errors of observation, efficiency here may be defined more broadly in terms of the interval $I_e$ as: $I_e = \{0.95 \le e_j \le 1.0; j=1,2,...,N_1\}$, so that the 95% efficiency level is considered to be efficient. The proportion $p_1(t)$ may then represent the probability $e_j$ falling in the inteval [0.95, 1.0]. Three types of dynamic persistence measures may then be proposed. One is very similar to the transition probability model discussed before, e.g., the lienar Markovian scheme:

$$p_1(t) = \alpha_1 p_1(t-1) + \alpha_0 + \varepsilon(t) \qquad (3.5.9)$$

where $\varepsilon(t)$ is the disturbance term. Given the observed data for $\{p_1(t), t=1,2,...,T\}$ the method of OLS (ordinary least squares) can be applied to estimate the parameters $(\alpha_1, \alpha_0)$, provided the error term $\varepsilon(t)$ satisfies the usual conditions of a white noise process. For convergence of the Markovian process the condition $0 \le \alpha_1 \le 1$ is also needed. Under these regularity conditions which are likely to be satisfied in most applications, one may define the dynamic efficiency process to be *persistent*, if the estimated value $\hat{\alpha}_1 \ge 0.95$. Highest persistence implies of course that $\hat{\alpha}_1$ is close to unity.

The disadvantage of the Markovian probability scheme (3.5.9) is that the specific efficiency levels, e.g., belonging to the interval $I_e$ are not considered at all. Next two persistence measures remedy this shortcoming. Consider now the efficiency levels $e_j(t)$ belonging to the interval $I_e$ only. We may decompose the series $\{e_j(t); 0 \le t \le T\}$ in terms of the conditional mean $\mu_t$ and conditional variance $\sigma_t^2$ as:

$$e_j(t) = E_{t-1}(e_j(t)) + \varepsilon_t ; \sigma_t^2 = E_{t-1}(\varepsilon_t^2) \qquad (3.5.10)$$

where $\mu_t = E_{t-1}(e_j(t))$ is the conditional expectation or mean and $\sigma_t^2$ the conditional variance in period t and both $\mu_t$ and $\sigma_t^2$ depend on the information set available up to period t. A model of *persistence in the mean* may then be formulated as

$$\mu_t = \beta_0 + \beta_1 \mu_{t-1} + error \qquad (3.5.11)$$

Note that the efficiency time sries $\{e_j(t)\}$ computed from the DEA models are only neeed here to estimate the conditional means $\hat{\mu}_t$ by following a moving average procedure. Moreover the residual error term in equation (3.5.11) also needs to be adjusted for serial correlation by following the Cochrane-Orcutt procedure; otherwise the estimates of the persistency parameter $\beta_1$ may be biased. Note that one has to impose also the condition $0 \le \beta_1 \le 1$ for convergence of the mean efficiency process. A significant estimated value $\hat{\beta}_1 \ge 0.95$ would then imply strong persistence, whereas $\hat{\beta}_1 < 0.50$ woud imply little or no persistence. Since the error term $\varepsilon_t = e_j(t) - E_{t-1}(e_j(t))$ may be viewed as shocks to the mean, the conditional variance $\sigma_t^2$ defiend in (3.5.10) may be used as a measure of volatility of shocks to the mean. If this volatility is large, the persistence of the mean efficiency process in (3.5.11) measured by a value of $\beta_1$ close to one may be less reliable. Hence the persistence of the volatility process needs to be modeled. A linear persistence model of volatility is given by

$$\sigma_t^2 = \gamma_0 + \gamma_1 \sigma_{t-1}^2 + \text{error} \tag{3.5.12}$$

For the linear volatility process model (3.5.12), volatility is persistent if $\gamma_1$ gends to a value close to unity or above. The so-called unit root or Arch (autoregression conditional heteroscedastic) models use this type of model frequently to characterize the nonstationarity and volatility of the financial markets. In this case the model specifies the conditional variance process as

$$\sigma_t^2 = \alpha_0 + \beta_1 \sigma_{t-1}^2 + \beta_2 \varepsilon_{t-1}^2 \tag{3.5.13}$$

where the error component $\varepsilon_{t-1}$ is estimated from equation (3.5.10).

Two reasons may be mentioned why the variance process (3.5.12) is important here. First of all, Farrell and Fieldhouse (1962) in their empirical applications to agricultural data found heteroscedasticity and skewness to be a pervasive feature of the empirical efficiency distribution. The Arch models (3.5.13) provide a simple framework for analyzing the persistence of heterogeneity in the output data due to heteroscedasticity. A second reason is that this specification allows a direct test of the *mean reversion hypothesis* of the efficiency distribution of firms. This hypothesis states that the higher deciles of the efficiency distribution generate smaller increases in efficiency than the lower deciles. In other words it applies the so-called law of proportionate effect which yields what is known as the Galtonian regression towards the mean.

A few comments are in order on the three types of persistence measures developed here in the framework of Farrell models. The transition probability model (3.5.9) may be solved to yield the time path of the proportion of the efficient firms as:

$$p_1(t) = \left[ p_1(0) - \frac{\alpha_0}{1 - \alpha_1} \right] \alpha_1^t + \frac{\alpha_0}{1 - \alpha_1} \qquad (3.5.14)$$

where the error term $\varepsilon(t)$ is set equal to zero. Here the parameter $(1-\alpha_1)$ measures the speed of adjustment since

$$\Delta p_1(t) = p_1(t) - p_1(t-1) = \alpha_0 - (1 - \alpha_1) p_1(t-1)$$

and this measure $(1-\alpha_1)$ indicates how quickly the proportion $p_1(t)$ of efficient firms reaches its long run equilibrium value $\bar{p}_1 = \alpha_0 (1 - \alpha_1)^{-1}$. When $\alpha_1$ is large, the proportion of efficient firms adjusts slowly to its permanent long run level. If, on the other hand, the value of $\alpha_1$ is small, the proportion of efficient firms is observed to be in long run equilibrium or steady state most of the time, except for purely random displacements. However it needs to be pointed out that Farrell cautioned against the use of this proportional measure $p_1(t)$ in comparing several dissimilar industries. This is because this measure does not fully reflect the extent to which the best practice in one industry compares with the best practice elsewhere. Hence this measure needs to be supplemented by mean persistence also.

Secondly, the mean persistence model (3.5.11) which can be generalized to second and higher order lags also, is comparable to total factor productivity growth or the Solow productivity residual applied by Solow (1957) and others. Recently Hall (1990) has estimated for U.S. manufacturing industries that this residual has procyclical fluctuations over time. The error term in the linear equation (3.5.11) could thus be empirically tested to see if there is any systematic fluctuations in $\mu_t$. For example assume that the error term $v_t$ in (3.5.11) follows a first order autocorrelated process:

$$v_t = \lambda v_{t-1} + w_t, \quad 0 < \lambda < 1 \qquad (3.5.15)$$

where $w_t$ is identically and independently distributed with a zero mean and constant variance. On combining (3.5.15) with (3.5.11) one obtains a second order model as follows:

$$\mu_t = \beta_0 (1 - \lambda) + \beta_1 \mu_{t-1} + \lambda \beta_1 \mu_{t-2} + w_t \qquad (3.5.16)$$
$$= b_0 + b_1 \mu_{t-1} + b_2 \mu_{t-2} + w_t$$

where $b_0 = \beta_0 (1 - \lambda), b_1 = \beta_1$ and $b_2 = \lambda \beta_1$. Clearly if the roots of the quadratic characteristic equation underlying the dynamic system (3.5.16) are complex, then this would indicate cyclical oscillations in the mean Farrell efficiency over time.

Finally, the volatility persistence model (3.5.12) in its second order form

$$\sigma_t^2 = \gamma_0 + \gamma_1 \sigma_{t-1}^2 + \gamma_2 \sigma_{t-2}^2 + \zeta_t$$

may be used to test volatility clustering, even when the error term $\zeta_t$ is a white noise process, i.e., has mean zero and fixed variance. The ratio $\sigma_t/\mu_t$ which measures the coefficient of variation may be usd to profile the average fluctuation of efficiency over time. Also the cyclical oscillations in the heteroscedasticity parameter $\sigma_t^2$ may be directly tested from the characteristic roots of the underlying characteristic equation.

Now we consider two empirical applications illustrating some of the measures of dynamic persistence in efficiency discussed earlier. The first application is in the public sector comprising school districts, while the second is in the private sector involving international air transport. In both cases, the inputs are very diverse and hence proxy variables are suitably utilized as explanatory variables for the production frontiers.

The first application utilizes the input output data in logarithmic units for selected public elementary school districts in California for eleven years 1977-88. Statistics of enrollment, average teacher salary, standardized test scores of students are all obtained from the published official statistics and reports. Out of a larger set of 35 school districts, 25 are selected in three contiguous counties of Santa Barbara, Ventura and San Luis Obispo on the basis of considerations of homogeneity of the student population and similarity in the average class size patterns.

The basic issue is to compare the productivity of academic units comprising diferent school districts. Several questions arise in this context. Are the inuts to the educational system being administered to maximize the quality of resulting outputs? Can the academic goals of the educational system be achieved within the current budgetary restraints? To what extent is the performance or quality of the educational system going to be lowered as a result of new sets of budgetary cuts? The search for answers to these and other questions of this type is hampered by the lack of acceptable measures of productivity and efficiency. Current methodology based on either all-purpose single-measure output-per unit of input indices (e.g., cost per student) or more sophisticated multivariate regression techniques have been found wanting, primarily due to the restrictive assumptions for their use and to the lack of inter-firm comparisons. Farrell's measure as generalized in DEA models provides a more convenient tool from this perspective and in the last few years it has emerged as a promising tool of generating performance efficiency mesures, which are independent of predetermined or arbitrary weights. Since our efficiency comparisons involved elementary school districts only, the output variable has to reflect the performance of pupils in standardized test scores. For the inputs we had to adopt a set of explanatory variables as the input proxies, which may explain the output performance. On the basis of the standardized test scores in reading, writing, spelling and mathematics, a composite weighted score is defined as average output y(t). We chose equal weights for the component scores since earlier results showed it to be better in terms of explanatory power. As input variables we had a choice of eight variables, of which the following four are selected in the Farrell LP model (3.5.3): $x_1(t)$ = average instructional expenditures, $x_2(t)$ = diversity index measured by the proportion of minority enrollment, $x_3(t)$ = average class size and $x_4(t)$ = average tax base of the district. Of these input variables the first is most important in the sense of knowledge-based human capital and the second reflects a growing trend for the state economy.

The data set for the whole period is summarized in Table 3.5.1 along with some statistics for the efficiency ratio. The estimates of the productoin frontier parameter $\beta_i^*(t)$ which are reported in Table 3.5.2 show that changes in $\beta_1^*(t)$ are more prominent and this is followed by $\beta_2^*(t)$ and $\beta_3^*(t)$. Over the last decase the state economy of California has been hit hard by budget cuts in the educational sector and also by increases in minority enrollment from students with a Spanish-speaking background.

Three major features are borne out by the estimates of persistence reported in Tables 3.5.3 and 3.5.4. First of all, the higher deciles in productivity ranking have experienced a steady persistence of efficiency with very little switch-over. For lower deciles, i.e., those units which are lower in the productive efficiency scale, this persistence is not upheld. This is clearly born out by the values of $\hat{\lambda}_i$ estimated from equation (3.5.7), i.e., for the $10^{th}$ decile it has a vlaue of 0.91 whereas for the first decile it is 0.12. Secondly, the mean efficiency model shows a remarkable persistence of efficiency over time. However this is also followed by a high persistence of the conditional variance of efficiency. Note that the estimate of $\alpha_1$ from model (3.5.9) is very high, i.e., 0.98 for the efficient units but the parameter $\gamma_1$ of variance persistence is also very high (i.e., $\hat{\gamma}_1$ = 0.89). This has two economic implications. One is that the mean efficiency is very sensitive to variance shocks due to sudden or large budget cuts and the heterogeneity in the school system is a significant contributing factor to the overall efficiency of school districts. The influence of the tax base in a school district also acquires a special importance. Although we have not reported the estimates, the effects of size are also important in the efficiency persistence process. For example a moderately large school district has a high efficiency profile over time, than either the very small or very large school districts.

The second application involves the airlines panel data set analyzed before by Sengupta (1995a). Here output is measured by ton-kilometers performed and the three inputs utilized are labor, fuel and capacity. Labor is measured by the number of employees, fuel by the expenditure in U.S. dollars and capacity by ton-kilometers available. Capacity is used as a proxy for the capital input. In our empirical applications all the inputs and output are measured in logarithmic terms, so that the production frontier is of a Cobb-Douglas form.

The annual data set for 11 Latin American airlines and 6 U.S. airlines over the period 1981-88 has been analyzed by Sengupta (1995b) before for estimating the production frontier. Here we present two sets of results in Tables 3.5.5 and 3.5.6 for five Latin airlines, where the DEA models are applied each year. It is clear from Table 3.5.5 that the efficiency persistence is uniformly maintained for only one airline, i.e., Avianca, whereas Peruvian Airlines remains inefficient all throughout. As in the first application, the persistence of efficiency for the efficient units is associated with volatility clustering also. This may perhaps provide a clue why Farrell in his study of British agricultural farms found the efficiency distribution to be highly skewed and heteroscedastic.

| Inputs | Outputs | Variance | | Efficiency |
|--------|---------|----------|--------|-----------|
| (log units) | (log units) | Inputs | Outputs | ratio (e) |
| $x_1$: 9.62 | y1 (reading): 4.179 | 0.017 | 0.018 | average: 0.92 |
| $x_2$: 2.960 | $y_2$ (writing): 4.179 | 1.040 | 0.015 | maximum: 1.00 |
| $x_3$: 3.237 | $y_3$ (spelling): 4.165 | 0.018 | 0.008 | maximum: 0.80 |
| $x_4$: 9.638 | $y_4$ (math): 4.071 | 12.185 | 0.020 | |
| | y: 4.153 | | 0.015 | |

Table 3.5.1. Characteristics of the input output data set (average over 1977-88)

| | 1977-78 | 1979-80 | 1981-82 | 1983-84 | 1985-86 | 1987-88 |
|--|---------|---------|---------|---------|---------|---------|
| $\beta_0^*(t)$ | -0.181 | -1.701 | -1.21 | -0.077 | -1.478 | -1.328 |
| $\beta_1^*(t)$ | 0.232 | 0.245 | 0.249 | 0.261 | 0.281 | 0.301 |
| $\beta_2^*(t)$ | 0.524 | 0.521 | 0.517 | 0.510 | 0.504 | 0.498 |
| $\beta_3^*(t)$ | 0.626 | 0.628 | 0.630 | 0.635 | 0.637 | 0.639 |
| $\beta_4^*$ | 0.180 | 0.178 | 0.171 | 0.164 | 0.161 | 0.151 |
| $y^*(t)$ | 4.243 | 4.261 | 4.296 | 4.383 | 4.441 | 4.413 |
| Percent efficient+ | 20 | 21 | 21 | 23 | 25 | 25 |

Table 3.5.2. Estimates of the dynamic production frontier for 2-year intervals

+Efficiency is measured by any value of $e_j \geq 0.95$, j=1,2,...,25.

| Decile | N | $\hat{\lambda}_j$ | Standard Error | $R^2$ | DW statistic |
|--------|---|-------------------|----------------|-------|--------------|
| 1 | 3 | 0.12 | 0.05 | 0.98 | 1.92 |
| 2 | 4 | 0.35 | 0.11 | 0.96 | 1.79 |
| 9 | 5 | 0.85 | 0.18 | 0.97 | 2.01 |
| 10 | 6 | 0.91 | 0.20 | 0.99 | 2.12 |

Table 3.5.3. Estimates of the transition probability parameters ($\lambda$j) for select deciles

| Model Type | Parameter Estimate | Standard Error | $R_2$ | DW |
|------------|--------------------|----------------|-------|-----|
| model (9) | $\hat{\alpha}_1 = 0.98$ | 0.28 | 0.994 | 2.03 |
| model (11) | $\hat{\beta}_1 = 0.97$ | 0.30 | 0.997 | 1.05 |
| model (12) | $\gamma_1 = 0.89$ | 0.29 | 0.972 | 2.01 |

Table 3.5.4. Estimates of the Markovian parameters $\alpha_1, \beta_1, \gamma_1$ for the efficien units

($e_j \geq 0.95$) only)

| Airlines | 1981 | 1982 | 1983 | 1984 | 1985 | 1986 | 1987 |
|----------|------|------|------|------|------|------|------|
| 1. LADE (E=25%) | E | E | N | N | N | N | N |
| 2. VARIG (E=62.5%) | E | E | E | E | E | N | N |
| 3. CRUZ (E=75%) | N | N | E | E | E | E | E |
| 4. AVIA (E=100%) | E | E | E | E | E | E | E |
| 5. PERU (E=0%) | N | N | N | N | N | N | N |

Table 3.5.5. Efficiency pesistence over time

Note: E and N denote efficient and nonefficient respectively.

| Model | Estimate | Standard error | $R^2$ | DW |
|-------|----------|----------------|-------|-----|
| model (3.5.9) | $\hat{\alpha}_1 = 0.98$ | 0.25 | 0.995 | 2.20 |
| model (3.5.11) | $\hat{\beta}_1 = 0.98$ | 0.27 | 0.989 | 1.98 |
| model (3.5.12) | $\hat{\gamma}_1 = 0.94$ | 0.18 | 0.985 | 2.14 |

Table 3.5.6. Estimates of the Markovian parameters $\alpha_1$, $\beta_1$, $\gamma_1$ for the efficient airlines

$(e_j \geq 0.95)$ only

# 4. Stochastic Efficiency Analysis

From an operational standpoint the stochastic aspects of efficiency analysis in DEA models arise in four different phases: (a) the characterization phase, where stochastic variations in input output data affect the efficiency frontier, (b) the control phase, where allocative efficiency is considered and the agent chooses optimal inputs and outputs subject to stochastic market prices, (c) the post-optimal phase in DEA models where the residuals or slacks from the efficiency frontier are analyzed in terms of their statistical distribution and finally (d) the estimation phase where a parametric form of the frontier is specified and an econometric method applied to estimate the stochastic frontier. In the efficiency literature the four phases are usually termed as: the stochastic data problem, the allocative efficiency analysis under price risk, the efficiency distribution approach and the stochastic frontier analysis (SFA). We discuss in this chapter the salient features of recent developments in these four phases of stochastic efficiency analysis.

## 4.1    Stochastic Data Problem

The stochastic variations of input output data may be incorporated into the efficiency characterization of a DEA model in two operational ways. One is to distinguish the data into two parts: a systematic part and an unsystematic part and then apply the DEA measure of efficiency on the basis of the systematic part of the input and output data. This is frequently done in portfolio theory where market returns have stochastic variations and the capital asset pricing model allows for a trade-off between mean return or yield against risk measured by standard deviations. The second method is to build stochastic data variations into the linear programming (LP) formulations of the DEA model and generate an efficiency frontier in terms of the resulting stochastic program. This section considers these two approaches in a statistical framework.

Consider the case of one output ($y_j$) and m inputs denoted by the vector $X_j$ where $j=1,2,...,N$ denotes a specific unit or DMU. For testing the relative efficiency of a reference unit $DMU_k$ one sets up the LP model

$$\text{Min } g_k = \beta_0 + \beta'X_k$$
$$\text{s.t.} \quad \beta_0 + \beta'X_j \geq y_j; j \in I_N = \{1,2,...,N\} \tag{4.1.1}$$
$$\beta \geq 0; \beta_0 \text{ free in sign}$$

where prime denotes the transpose of a vector. If the k-th firm or $DMU_k$ is technically efficient, then $y_k = y_k^*$ where $y_k^* = \beta_0^* + \beta *' X_k$ is the production frontier. The errors in the production function may be written as

$$\varepsilon_j = y_j - \beta_0 - \beta' X_j ; j \in I_N$$

where

$$\varepsilon_j = v_j - u_j; v_j = \text{statistical noise}, u_j \geq 0 \tag{4.1.2}$$

The nonnegative error term $u_j$ represents technical inefficiency, whereas the $v_j$ are identically and independently distributed random variables which represent those effects which cannot be controlled by the firms such as quality, the uncertainty about input supply and the effects of left-out explanatory variables.

The method proposed here is to decompose the composite error $\varepsilon_j$ in (4.1.2) into two parts: the systematic and the unsystematic and then apply a Box-Cox type transformation to the input output data so that nonnormality of the distribution of the composite error $\varepsilon_j$ can be transformed to approximate normality. Let $d_j = (X_j, y_j)$ belonging to the data set $D = \{d_j, j \in I_N\}$ denote the input output data for $DMU_j$ such that it is decomposed into a systematic part $(d_j^s)$ and an unsystematic part $(d_j^u)$ i.e., $d_j = d_j^s + d_j^u, j \in I_N$. The standard methods of statistical filtering may be employed so as to determine the systematic parts of the input output data. On using this systematic data we can determine by the DEA model (4.1.1) the systematic production frontier. For time series data with nonstationary trends we may determine the systematic part by fitting a suitable polynomial function of time. For example assume that the first difference $\Delta y_{jt}$ of the output series makes it stationary, then the systematic component $y_{jt}^s$ can be specified by the first degree fitted polynomial in time i.e., $y_{jt}^s = a_{0j} + a_{1j}t$. In the general case a p-th degree polynomial in time may have to be fitted to the observed input output time series.

To handle nonnormality of the error term $\varepsilon_j$ we may apply the method of Box-Cox transformation to the data points, $d_j$. Thus if only output is nonnormally distributed, the method constructs a transformation of the dependent variable y such that it linearizes the model, corrects for heteroscedasticity and normalizes the distribution of errors. The type of transformation would vary from one data set to another depending on the nature of the nonnormal distribution, see e.g., Carroll and Ruppert (1988) for the details of such transformations. For example one transformation which is frequently useful in empirical studies is of the form

$$z_j^{(\lambda)} = \begin{cases} z_j^\lambda, \lambda \neq 0 \\ \ell n z_j, \lambda = 0 \end{cases}$$

where $z_j$ is any input or output.

The rationale behind using the systematic components of the data set to determine the DEA efficiency frontier may be traced to the adoption of quality control techniques in industrial production. Thus when input variability is the primary cause of output variability, one could model the input data one step further e.g.,

$$x_{ij} = \mu_i + d_i z_j + \varepsilon_{ij} \qquad (4.1.3)$$

where $\mu_i$ is the mean of the i-th population (row effect), $z_j$ is a proxy variable representing other factors which require this input (column effect) and $\varepsilon_{ij}$ is the random error. For cross section data the mean effect $\mu_i$ can be estimated by the sample mean, whereas for panel data one could further decompose the mean level as $\mu_i = \bar{\mu}_i + L_i$, where $L_i = \mu_i - \bar{\mu}_i$ now represents the distance of the mean input i from its grand mean $\bar{\mu}_i$. It is important to note that the quantity $\mu_i$ being the population mean for input i is a constant when viewed at a fixed time point t. But when viewed from the perspective of a time horizon it is a random variable with variance $\sigma_L^2$. The other variables denoted by $z_j$ in (4.1.3) may represent measures of technological change, or cumulative output as a proxy for technological progress. On using the estimated $\hat{x}_{ij}$ from (4.1.3) one could set up an input-oriented DEA model as a convex hull:

$$\text{Min } \theta$$
$$\text{s.t.} \quad \sum_j \hat{x}_{ij} \lambda_j \leq \theta \hat{x}_{ik}; \sum_j y_{rj} \lambda_j \geq y_{rk} \qquad (4.1.4)$$
$$\sum_j \lambda_j = 1, \lambda_j \geq 0; j \in I_N$$

One may also transform the output variables by their systematic components $\hat{y}_{rj}$ also. The systematic measures of technical efficiency would then be given by the optimal value $\theta^* = 1.0$ in (4.1.4) with all slack variables zero. Clearly the systematic measures of technical efficiency would be more stable in the sense of lower variance. This two-step method involving separation of the systematic and the unsystematic components and then using the systematic components only is widely adopted in practical applications of the modern theory of quality control. Sengupta (1996) has successfully applied this technique for DEA models in respect of the international airlines data over the period 1981-88 showing that it produces a reliable estimate of the scale economies and also dynamic efficiency in air transport.

Incorporating the stochastic variations of data $d_j$ by transforming the original DEA model has followed two procedures. One is the method of chance-constrained programming where the input output constraints are assigned tolerance measures. For instance the input-oriented model gets transformed as

$$\text{Min } \theta$$

$$\text{s.t.} \quad \text{Prob}\left[ \sum_{j=1}^{N} Y_j\lambda_j \geq Y_k \right] \geq 0.95 \tag{4.1.5}$$

$$\sum_{j} X_j\lambda_j \leq \theta X_k; \lambda_k \geq 0; \ \theta \text{ free in sign}$$

where $\alpha = 0.95$ for example may be preassigned implying that the observed outputs must not exceed best practice outputs more than 5% of the time. The deterministic equivalent of the above problem is nonlinear in the case when the outputs are normally distributed.

The second method is applicable when the panel data are available and the input or output fluctuations can be measured by their estimated variances. Thus Murthi, Choi and Desai (1977) have applied this method to evaluate the performance of mutual funds in terms of the DEA portfolio index (DPEI). To test the efficiency of fund k they set up the ratio model

$$\text{Max DPEI}_k = y_k \left( \sum_{i=1}^{m} w_i x_{ik} + v\sigma_k \right)^{-1}$$

$$\text{s.t.} \quad y_k \leq \sum_{i=1}^{m} w_i x_{ij} + v\sigma_j \leq 1; j \in I_N \tag{4.1.6}$$

$$w_i, v \geq \varepsilon, \ \varepsilon = \text{fixed positive value}$$

where $w_i$, $v$ are positive weights, $x_{ij}$ are the value of the ith transaction costs (such as turnover, expense ratio, loads, etc.) for the jth fund j and $\sigma_j$ is the risk of fund j measured by the estimated standard deviation of return. The above can be transformed to a linear model. Clearly the impact of the risk variable $\sigma_j$ on the efficiency frontier can be directly evaluated in this framework.

## 4.2    Efficiency Frontier Under Price Uncertainty

Price or allocative efficiency measures the firm's success in choosing an optimal set of inputs with a given set of input prices. But market prices even in competitive frameworks are subject to stochastic variations due to various factors such as weather, bottlenecks in supply and even seasonal factors. We consider in this section the implications of input price fluctuations on the optimal inputs chosen by the firms. Let q be the input price vector with mean $\bar{q}$ and variance covariance vector V. Assume that each DMU or firm is risk averse in the sense that it minimizes a risk adjusted loss function as

$$\phi = \bar{q}'x + (w/2)x'Vx \qquad (4.2.1)$$

To test the overall efficiency of $DMU_k$ one minimizes the loss function $\phi$ in (4.2.1) subject to the constraints

$$\sum_{j=1}^{N} X_j\lambda_j \le x; \sum Y_j\lambda_j \ge Y_k; \sum \lambda_j = 1 \qquad (4.2.2)$$

$$\lambda, x \ge 0$$

where w is a nonnegative weight representing the degree of risk aversion, higher (lower) values for higher (lower) aversion. This type of representation is quite standard in stochastic economics, e.g., portfolio theory in investment models. Note that the risk adjusted cost function $L = L(c)$, $c = q'x$ admits of the following interpretation: if each firm's loss function $\phi(c)$ of the exponential form:

$$L = L(c) = \exp(wc), w > 0 \qquad (4.2.3)$$

with a constant rate (w) of absolute risk aversion and the cost c is normally distributed as $N(\bar{q}'x, x'Vx)$, then minimizing the expected value $E\{L(c)\}$ of this loss function is equivalent to minimizing the risk adjusted cost function $\phi$ as defined in (4.2.1.). This formulation of the risk adjusted cost functional $\phi$ has several advantages in applied statistical decision theory. First of all, if cost $c = q'x$ is not normally distributed but subject to a specific probability distribution such as truncated normal, one can expand the loss function (4.2.3) up to quadratic terms and derive

$$E\{L(c)\} \simeq 1 + wE(c) + (w^2/2)\{var\, c + (Ec)^2\}$$

Then the risk adjusted cost function becomes

$$\phi \simeq 1 + w[\bar{q}'x + (w/2)\{x'Vx + (\bar{q}'x)^2\}]$$

which is very similar to the specification (4.2.2). Secondly, one could apply the so-called safety-first rule of stochastic programming so as to obtain more robust optimal solutions. Let $C_\alpha(x)$ denote the $\alpha$-quantile of the stochastic cost function $c = q'x$, i.e., $Prob[c(x) = q'x \le C_\alpha(x)] = \alpha$, $0 \le \alpha \le 1$. If $c(x)$ is normally distributed then it follows

$$C_\alpha(x) = \overline{q}'x + \theta_\alpha (x'Vx)^{1/2} \tag{4.2.4}$$

where $\theta_\alpha = \Phi^{-1}(\alpha)$, $\Phi(\cdot)$ being the cumulative distribution function of a unit normal variate. In the general case the transformed DEA model becomes

$$\underset{x,\lambda}{\text{Min}} \, C_\alpha(x) \text{ s.t. the constraints of (4.2.2)} \tag{4.2.5}$$

Clearly this defines a nonlinear programming model if the constant $\theta_\alpha$ is different from zero. Let $x^{**} = x_\alpha^{**}$ and $\lambda^{**} = \lambda_\alpha^{**}$ be the optimal solutions in this nonlinear model for a fixed $\alpha$. Then the efficiency measure for the reference unit $DMU_k$ can be measured by

$$E_2(k) = C_\alpha^{**}(x)/C_\alpha(x) \text{ for a fixed } \alpha \tag{4.2.6}$$

where $C^{**}(x) = C(x_\alpha^{**})$. This can be compared with the deterministic measure

$$E_1(k) = c_k^* / c_k$$

where $c_k = q'X_k$ and $c_k^* = q'x^*$ are the observed and optimal costs. The normal distribution case in (4.2.1) yields the third measure:

$$E_3(k) = \phi^*/\phi \text{ for a fixed } w.$$

Since $\theta_\alpha \geq 0$ for $\alpha \geq 0.50$ it is clear that the second term on the right hand side of (4.2.4) can be interpreted as a penalty for uncertainty. For the case $\alpha = 0.50$, the criterion function (4.2.4) reduces to the Laplace criterion of minimizing the expected cost function. In case $\alpha$ tends to 1.0 we get the maximin criterion of Wald. Thus various types of risk sensitive attitudes can be built into this framework.

One limitation of the DEA model above is that it does not distinguish between current and capital inputs, although most production processes involve both, i.e., current output depends on both current and capital inputs. Let $K_j = (k_{ij})$ and $I_j = (\Delta k_{ij})$ denote the vectors of capital inputs and investments ($i=1,2,\ldots,m_1$; $j=1,2,\ldots,N$) where investment $I_{ij} = \Delta k_{ij}$ is the incremental capital stock. Then a more generalized version of the DEA model may be specified as follows:

$$\text{Min } \theta$$

$$\text{s.t.} \quad \sum_{j=1}^{N} X_j\lambda_j \leq \theta X_k ; \sum_{j=1}^{N} K_j\lambda_k \leq \theta K_k$$

$$\sum_{j=1}^{N} I_j \lambda_j \leq \theta I_k ; \sum_{j=1}^{N} Y_j \lambda_j \geq Y_k \tag{4.2.7}$$

$$\lambda'e = 1, \lambda_j \geq 0; j = 1, ..., N$$

By duality this yields a production function as

$$\beta'X_j + \psi'K_j + \gamma'I_j \geq \alpha'Y_j + \mu$$

Here if the reference unit $DMU_k$ is technically efficient then it must satisfy the production frontier condition as:

$$\beta*'X_k + \psi*'K_k + \gamma*'I_k = \alpha*'Y_k + \mu* \tag{4.2.8}$$

This specification has several flexible features. The first is that efficiency now requires that $\theta* = 1.0$ with equality signs holding for the four sets of constraints (4.2.7). This implies full utilization of capacity $(K_k)$ and the new investment inputs $(I_k)$ for the reference unit. Since full capacity utilization is normally considered to be an engineering measure of efficiency for machine utilization systems in manufacturing, this specification is very helpful. The second feature is that if the reference unit is not efficient, then the specific sources of inefficiency can be identified, e.g., if it is due to the lack of utilization of current capital or investment inputs or all of them. Finally, the overall efficiency analysis can now be directly incorporated by replacing the observed input vectors $X_k, K_k, I_k$ with the decision vectors x, K, I for which the observed market prices are q, r and h respectively. The model then becomes

$$\text{Min } c = q'x + r'K + h'I$$
$$\text{s.t.} \quad \sum_j X_j \lambda_j \leq x; \sum_j K_j \lambda_j \leq K; \sum_j I_j \lambda_j \leq I \tag{4.2.9}$$
$$\sum_j Y_j \lambda_j \geq Y_k ; \lambda'e = 1, \lambda_j \geq 0; j = 1, ..., N$$

Once again the stochastic counterpart of this model (4.2.9) can be specified along the lines specified before.

For example the model analogous (4.2.1) may be formulated as:

$$\text{Min } \phi = \bar{q}'x + \bar{r}'K + \bar{h}'I + (w/2)[x'V_q x + K'V_r K + I'V_h I + \text{cov.}]$$
$$\text{s.t.} \quad \text{the constraints of (4.2.9)}$$

where the cov. term denotes the covariance of the three component costs of $q'x, r'K$ and $h'I$. It is clear that the three sets of market price data (q,r,h) must be sufficiently large so as to obtain reliable estimates of the mean and covariance parameters. In case of normal distribution the reliable estimates of these parameters may be obtained through the sufficient statistics, provided the data set is large. But for nonnormal cases, problems of estimation risk exist and they may affect the efficiency measurement in the stochastic case.

Next consider dynamic efficiency in the intertemporal sense. This type of dynamic efficiency arises in most modern production processes, because they take several time periods to adjust their inputs and outputs to the desired levels. Hence the static DEA models that are exclusively based on current inputs frequently generate biases in efficiency measurement due to several factors. First of all, capital inputs and technical innovation generate output profiles over time, and the utilization of production capacity created by new investment may not be spread uniformly over several time periods. Secondly, the firms or DMUs tend to learn and adjust over time once they acquire relevant information about the sources of inefficiency. When new technology is the source of productivity gain, the firms may take several periods to learn about the new technology and adopt it fully. Thus Norsworthy and Jang (1992) have recently found from empirical data that the impact of such learning on industrial productivity growth is considerable and spread over several periods in the modern technology-intensive industries in Japan and the U.S., e.g., semiconductor, microelectronics and communications. Finally, when the input and output processes are subject to price fluctuations, the risk averse firms usually adopt a cautious policy in changing to a more efficient technology or production system.

Consider a dynamic intertemporal extension of the overall efficiency model in a simplified framework. We may assume without loss of generality that there is only one capital input $K_j$ for each firm or $DMU_j$ and m current inputs ($X_j$) and s outputs ($Y_j$) as before. The reference unit k has to optimally decide on the level K of capital input, which is subject to the following equation of motion:

$$\dot{K} = I - \delta K \qquad\qquad (4.2.10)$$

where $\dot{K} = dK/dt$ is net investment, as the rate of change of capital stock, $\delta$ is the constant rate of depreciation (assumed to be known), and I is gross investment. The overall efficiency model is then of the form:

$$\text{Min } J = \int_0^\infty \exp(-\rho t)[q'x(t) + c(I)]dt$$

$$\text{s.t.} \quad (4.2.10) \text{ and } \sum_{j=1}^{N} X_j \lambda_j \le x(t)$$

$$\sum_{j=1}^{N} Y_j \lambda_j \ge Y_k(t); \sum_{j=1}^{N} K_j \lambda_j \le K(t) \qquad (4.2.11)$$

$$e'\lambda = 1, \lambda \ge 0; x(t) \ge 0 .$$

The decision variables are $\lambda = \lambda(t)$, $x(t)$, $I(t)$ and $K(t)$ and the observed data are the input and output vectors $X = X_j(t)$, $Y = Y_j(t)$, the capital used by firm j $(K_j)$ where the reference unit is k. The term $c(I)$ is the investment cost function, which can be interpreted in two ways. One is that it represents the cost of purchased units of extra capital, i.e., $c(I) = hI(t)$, with h as the unit price. Also it may be interpreted as adjustment costs, i.e., $c(I) = c(K,\dot{K})$ so that it reflects the disequilibrium cost of changing from a given level of $K = K(t)$ to a target or desired level corresponding to future demand.

The efficiency model above minimizes a discounted stream of total input costs composed of current inputs and investment inputs generating adjustment costs, where $\rho$ is a positive rate of discount assumed to be known. With nonnegative input output vectors $(X_j, Y_j)$ the constraint set is compact and if the cost function $c(I)$ is convex, then an optimal trajectory exists and it can be characterized in terms of the Pontryagin maximum principle. Let H be the Hamiltonian function

$$H = \exp(-\rho t)[-q(t) - a_1 I - (a_2/2)I^2 + p(I - \delta K)$$
$$+s'(K - \sum_{j=1}^{N} K_j\lambda_j) + \beta'(x - \sum_j X_j\lambda_j) \qquad (4.2.12)$$
$$+ \alpha'(\sum_{j=1}^{N} Y_j\lambda_j - Y_k) + \mu(e'\lambda - 1)]$$

where the investment cost function is assumed to be a convex quadratic function. The optimal trajectory then satisfies the following necessary conditions:

(a)     There must exist a continuous function p(t) such that

$$\dot{K} = I - \delta K$$

with initially given $K(0) = K_0$                                              (4.2.13a)

and

$$\dot{p} = (\delta + \rho)p - s \quad \text{(perfect foresight condition)}$$

(b)     and at each moment of time t the cost function

$$c = q'x(t) + a_1 I + (a_2/2)I^2 - pI$$

is minimized with respect to $I(t)$, $K(t)$, $x(t)$ and $\lambda_j(t)$
subject to all the constraints of (4.2.2) except (4.2.10),                     (4.2.13b)

and the transversatility condition

$$\lim_{t \to \infty} \exp(-\rho t)p(t) = 0 \qquad (4.2.13c)$$

The second set of conditions yields by duality the following constraints to hold for each time point t:

$$\beta \le q; a_1 + a_2 I \ge p(t); s'K_j + \beta'X_j \ge \alpha'Y_j + \mu \qquad (4.2.14)$$

which is very similar to the dual constraint (4.2.7) of the static DEA model, except for the constraint on the control variable $I = I(t)$. Let asterisks indicate optimal values and the reference unit k be efficient over time. Then it must satisfy the following equalities:

$$\beta^* = q(t); a_1 + a_2 I^*(t) = p^*(t)$$
$$s^{*'} K_k + \beta^{*'} X_k = \alpha^{*'} Y_k + \mu^* \qquad (4.2.15)$$
$$\dot{K}^*(t) = I^*(t) - \delta K(t); \dot{p}^* = (\delta + \rho)p^*(t) - s^*(t)$$

Several interesting features of dynamic efficiency are characterized in this specification (4.2.15). First of all, the dynamic trajectory of optimal investment $\{I^*(t); 0 < 1 < \infty\}$ determines the optimal time path of capital $\{K^*(t), 0 < t < \infty\}$. On using these two trajectories at each time point t, the static model specified by (4.2.13b) determines the static efficiency condition. Secondly, the static cost of capital for the efficient $DMU_k$ is $s^*(t) K^*(t)$ with a positive $s^*(t)$, whereas the dynamic cost at time t is $[s^*(t) K^*(t) - \delta p^*(t)]$ which is less (more than the static cost if the dynamic shadow price $p^*(t)$ of capital accumulation is positive (negative). Clearly the static and dynamic efficiency measures are very different so long as the dynamic shadow price $p^*(t)$ is not equal to zero. Thirdly, the production frontier equation in (4.2.15) shows the presence of the capital input $K_k$ along with its shadow price $s^*$ at each time point t. In the static case this term is absent. Finally, the transversatility condition (4.2.13c) ensures that there is a steady state of the optimal trajectories, which corresponds to the long run level of dynamic efficiency. Convergence to the steady state level of efficiency can be analyzed in this framework by following the standard methods of optimal control theory; see, e.g., Sengupta (1994). For example the steady state values can be specified as follows:

$$\bar{I}^* = (\bar{p}^* - a_1)/a_2 \qquad (4.2.16)$$
$$\bar{K}^* = \bar{I}^*/\delta; \bar{p}^* = \bar{s}^*(\delta + \rho)^{-1}$$

This shows then that $\bar{p}*$ rises if either $\bar{s}*$ rises or $\rho$ falls, whereas a rising $\bar{p}*$ implies an increase in steady state investment $\bar{I}*$. Furthermore, if $p*(t)$ rises (falls) then $\dot{K}*(t)$ rises (falls) since

$$\dot{K}*(t) = (1/a_2)[p*(t) - a_1] - \delta K*(t) \qquad (4.2.17)$$

In case of stochastic variations in input prices we rewrite the objective functional of (4.2.13) as

$$\text{Min } \hat{J} = \int_0^\infty \exp(-\rho t)[\bar{q}'x(t) + (w_1/2)x'V_q x$$
$$+ \bar{c}(I) + (w_2/2)\text{ var } c(I)]dt \qquad (4.2.18)$$

where $w_1$ and $w_2$ are nonnegative weights and $\bar{c}(I)$, var $c(I)$ are the mean and variance of the adjustment cost function. In this case the problem becomes one of nonlinear programming and it reduces to a quadratic program if the var $c(I)$ function can be approximated by a quadratic term. Clearly in these cases the variance of input prices and the adjustment costs would affect the dynamic efficiency of the reference unit $DMU_k$. Thus in the dynamic case two efficiency measures can be suitably defined with respect to the models (4.2.2) and 4.2.9):

$$E_4(k) = J*/J; E_5(k) = \hat{J}*(w_1, w_2)/\hat{J}(w_1, w_2)$$

where the weights $w_1$, $w_2$ are assumed to be fixed.

Two general comments may be made about the allocative efficiency model under price uncertainty. One is the case where output prices p are also stochastic and both inputs (x) and outputs (y) are chosen as optimal control variables. In this case the objective function (4.2.1) is replaced by

$$\text{Max}\pi = \bar{p}'y - \bar{q}'x - (w/2)[y'V_p y + x'V_q x - 2y'V_{pq}x] \qquad (4.2.19)$$

and the vector $Y_k$ (4.2.2) is replaced by the control vector y. We thus obtain a stochastic profit efficiency frontier. Secondly, the objective function (4.2.19) may also be viewed more generally in terms of maximizing the expected value $EU(\tilde{\pi}; \tilde{p}, \tilde{q})$ of utility of profits $\tilde{\pi}$, when the prices $\tilde{p}, \tilde{q}$ follow a specified class of statistical distributions. The form of the utility function may then vary as is frequently done in portfolio investment theory.

An important issue here is to ask if there exists an optimal solution $(x*, y*)$ for a given class of utility functions $U(\tilde{\pi}; \tilde{p}, \tilde{q})$ such that it has first or second order stochastic dominance over any other feasible vector (x,y). The robustness of the optimal solution may then be statistically tested.

## 4.3     Efficiency Distribution Analysis

This section illustrates the efficiency distribution approach by two empirical applications. The first involves a cost frontier model, where total cost is viewed as a loglinear function of output and the input prices. The objective is to estimate the distribution of costs around the efficient output levels. The second involves measuring dynamic efficiency over time.

Consider for example an output-oriented model where each DMU has m inputs and n outputs and there are N units in the whole sample. Let X and Y be the input-output matrices of dimensions m by N and n by N respectively and let $X_k$, $Y_k$ be the input and output vectors for the reference unit k which is to be compared with other units.

$$\text{Min } gk = \beta'X_k$$
$$\text{s.t.} \quad \alpha'Y_k = 1; \beta'X \geq \alpha'Y$$
$$\alpha \geq 0, \beta \geq 0$$

Here prime denotes the transpose and sometimes the nonnegativity condition on $\alpha$, $\beta$ is replaced by $\alpha \geq \varepsilon e$, $\beta > \varepsilon e$ to assure nondegenerate solutions, where $\varepsilon$ is a positive non-Archimedean infinitesimal and e is a vector of unit elements. The dual of this problem is then easily derived as

$$\text{Max } z_k = \phi$$
$$\text{s.t.} \quad Y\lambda \geq \phi Y_k; e'\lambda = 1; \lambda \geq 0$$

Denote the optimal solutions by asterisks. Then it is well known in the DEA literature that the following two statements are equivalent:

(1)     A $DMU_k$ is efficient if and only if
        $\phi^* = 1.0$ and (b) all slacks are zero and

(2)     A $DMU_k$ is efficient if and only if $g_k^* = z_k^*$

For simplicity of exposition we may consider the case of one output with m inputs. In this case the condition $\alpha'Y_1 = 1$ drops out.

For any $DMU_j$ we denote its observed output by $y_j$ and the inputs by vector $X_j = (x_{ij})$, where $j \in I_N$ and $I_N = \{1,2,...,N\}$ is the index set of N units in the sample. To test the technical efficiency of the reference unit $DMU_k$, the DEA approach sets up the LP model for the kth unit as

$$\text{Min } g_k = \beta'X_k$$
$$\text{s.t.} \quad \beta'X_j \geq y_j; \beta \geq 0; j \in I_N \qquad (4.3.1)$$

Here prime denotes the transpose and the intercept term is subsumed here by interpreting one of the inputs to equal unity. Let $\beta^* = \beta^*(k)$ be the optimal solution and assume it for simplicity to be nondegenerate. Then define $y_k^* = \beta^*(k)'X_k$ as the optimal output associated with the production frontier. If $y_k^* > y_k$ at the optimal solution, then the reference unit $DMU_k$ is relatively inefficient, since its observed output $y_k$ is dominated by its potential output $y_k^*$. Thus efficiency holds if $y_k^* = y_k$ for the reference unit. Now by varying k in the objective function of (4.3.1) over all the N units, one could determine the subset of units say $N_1$ in number ($N_1 < N$), which is relatively efficient in the above sense. The remaining units $N_2 = N—N_1$ are then inefficient. Let $S_1$ and $S_2$ be these two subsets of the overall sample S, where $S_1$ contains the $N_1$ efficient units and $S_2$ the rest. The efficiency distribution approach analyzes the distribution of output in the efficient subset $S_1$ and compares it with that in subset $S_2$, which comprises the inefficient units. In efficiency literature, the statistical distribution of output has been analyzed by Farrell and Fieldhouse (1962), and more recently by Tulkens and Eeckaut (1995). These studies however are mainly graphical in nature based on histogram estimates. They do not incorporate the various characteristics of the specific form of the empirical distribution, e.g., features of nonnormality or heteroscedasticity. Nor do they relate the efficiency distribution to the estimates of the productivity parameters such as $\beta$ in the production frontier relation:

$$y_j = \beta'X_j - \varepsilon_j, \varepsilon_j \geq 0, j \in S_1 \qquad (4.3.2)$$

when the empirical distribution of the residual error term ($\varepsilon_j$) is determined by the LP models in (4.3.1). In a static framework where the input output data are from cross-section observations, one could adopt two different ways of looking at the efficiency distribution approach. One is to look at the distribution of output belonging to the efficient subset $S_1$. The various moments of this distribution may then be used here to characterize the distribution. Also, this distribution can be compared with the other distribution for the inefficient units in $S_2$ and also the overall actual distribution of observed output. To what extent the efficient units can serve as a target set or a peer group may be evaluated here in terms of a suitable measure of distance between the two distributions. Secondly, one could set up a model of the form (4.3.2) with the restriction that $j \in S_1$ and the estimate the parameter vector $\beta$ by suitable statistical procedures. These procedures would incorporate in its estimation framework the empirical distribution of the error terms ($\varepsilon_j$) as determined in the first step of the LP formulation of the DEA models. For convenience of presentation we consider the second method first and then the first method.

Consider the subset $S_1$ of efficient units, where each $DMU_k$ satisfies the equality

$$y_k = y_k^* = \beta^{*\prime}(k)X_k, k \in S_1 \tag{4.3.3}$$

Clearly this generates $N_1$ vector parameters with $mN_1$ elements, though all elements may not be distinct.

By contrast if we are to estimate an econometric production frontier over the set $S_1$, we could use the specification (4.3.2) for $j \in S_1$ and the empirical distribution of the error term $(\varepsilon_j)$ to derive a single vector estimate of $\beta$ with m elements. The simplest procedure would adopt the following steps:

Step 1. In the first step one estimates the empirical frequency function of the error terms $\varepsilon_j = \beta^\prime X_j - y_j$ from the original DEA model. Here one can apply the method of moments in its simple or generalized form to estimate the probability density function $p(\varepsilon_j)$. This estimate can be further improved by the maximum likelihood (ML) method if so desired.

Step 2. Given the form of the empirical distribution $p(\varepsilon_j)$ of the error terms, the second step would apply the ML method to estimate the parameter vector $\beta$.

Step 3. Conditional on the estimated parameter vector $\hat{\beta}$, the third step derives the conditional moment estimates of the distribution of optimal output over the subset $S_1$.

Step 4. Finally, on using the mean $\hat{\mu}$ of the empirical probability density function $p(\varepsilon_j)$, one could also propose the method of corrected ordinary least squares (COLS), whereby we apply the least squares method to the transformed model

$$y_j = -\hat{\mu} + \beta^\prime X_j + u_j, u_j = \hat{\mu} - \varepsilon_j$$

This estimate of $\beta$ by the COLS method may then be compared with the ML method applied in Step 2 above. The comparison may be made either by a parametric procedure such as the chi-square test of goodness of fit, or a nonparametric procedure such as Kolmogorov-Smirnov test statistic.

Whereas the four steps above have the central focus on the estimation of the production or cost frontier from the data set $S_1$ containing the efficient units only, the other aspect of the efficiency distribution approach looks at the distribution of optimal output in a comparative sense. Two major issues here are the following:

First, how does this distribution compare with that of the inefficient units belonging to the set $S_2$? Comparisons based on the criterion of stochastic dominance of different orders may be very helpful here; second, how does one construct a "peer group" of efficient units belonging to the set $S_1$, so that the gap in performance between the inefficient units and the peer group can be quantitatively assessed? Sengupta (1990) has suggested a peer group criterion based on the concept of *modal efficiency* or core efficiency. A subset $S_M(p)$ of the efficient set $S_1$ is said to satisfy the modal efficiency condition, if each of its units turns out to be efficient in N LP models at least p percentage of times. Clearly p lies between $1/N$ and 1, since the lowest frequency of occurrence is $1/N$ and the highest is 100%. Let $p_1, p_2, ...$ be the relative frequencies of occurrence of efficiency, ordered as $p_1 \geq p_2 \geq p_3 \geq ... p_k > 0$. Then $p_1$ is the modal efficiency set $S_M(p_1)$. Next highest is the set $S_M(p_2)$ with the associated level of efficiency $p_2$. One can use then the set $S_M(p_1)$ or, a combination of $S_M(p_1)$ and $S_M(p_2)$ as the peer group, provided $p_2$ is very close to $p_1$ and $p_1$ exceeds 0.50, i.e., the units in set $S_M(p_1)$ is at least 50% of the time efficient in N LP models. Then one can evaluate the role of the modal efficiency set by running a regression of the observed output on the m inputs and a zero-one dummy variable $D_j$ defined as

$$y_j = \beta'X_j + \alpha D_j + u_j; u_j = \text{error with zero mean}$$

where (4.3.4)

$$D_j = \begin{cases} 1, & \text{if } j \in S_M(p_1) \\ 0, & \text{otherwise} \end{cases}$$

Sengupta (1989) has described in some detail the various applications of this dummy variable regression approach in the framework of data envelopment analysis. This formulation (4.3.4) helps to explain for an inefficient unit the efficiency gap $\varepsilon_j = \beta'X_j - y_j$ in terms of the distance from the peer group represented here by the dummy variable $D_j$. Since the statistical significance of the regression coefficient $\alpha$ in (4.3.4) may be tested here, it provides a direct measure of importance of the peer group for the inefficient units belonging to the subset $S_2$.

The efficiency distribution approach has two important dynamic implications depending on the form of dynamics entering into the specification. One arises when the LP model (4.3.1) is computed at each time point t where $t \in I_T = \{1, 2, ..., T\}$, i.e.,

$$\min g_{kt} = \beta'_t X_{kt}$$
$$\text{s.t.} \quad \varepsilon_{jt} = \beta'_t X_{jt} - y_{jt} \geq 0; \beta_t \geq 0 \quad (4.3.5)$$

Since the random variables $\{\varepsilon_{jt}\}$ are not normally distributed due to nonnegativity, it is more likely to have heteroscedasticity and other nonnormal features such as significant skewness and kurtosis. On using the empirical estimates of the random vector $\varepsilon_t$, one could test the persistence of efficiency over time. Furthermore one could test if there is any stability of efficiency over time, or there is volatility when the latter is measured by the conditional variance of $\hat{\varepsilon}_t$, given the available information up to time point t-1, see e.g., Sengupta (1996d,e).

The second specification assumes that the input output data are all nonstationary but their first differences are stationary. Thus the change in output $\Delta y_{jt} = y_{jt} - y_{j,t-1}$ is now due to two factors: the change in the input vector $\Delta X_{jt} = X_{jt} - X_{j,t-1}$ and the state of technology embodied in inputs $X_{j,t-1}$ at time t-1. The dynamic DEA model now takes the form:

$$\min g_{kt} = \lambda' \Delta X_{jt} + \theta \varepsilon_{k,t-1}$$
$$\text{s.t.} \qquad \gamma' \Delta X_{jt} + \theta \varepsilon_{j,t-1} \geq \Delta y_{jt} \qquad\qquad (4.3.6)$$
$$\varepsilon_{jt} = \beta' X_{jt} - y_{jt}$$
$$\beta \geq 0, 0 \leq \theta \leq 1, \gamma: \text{free in sign.}$$

Note that this model assumes for simplicity that the parameters $\gamma$, $\theta$ and $\beta$ are all constants and each programming model (4.3.6) solves for their optimal values. Although this appears to be nonlinear in form, it can be solved by an LP routine by rewriting the constraints as

$$\gamma' \Delta X_{jt} + \delta' X_{j,t-1} \geq \Delta y_{jt} + \theta y_{j,t-1}$$

where $\qquad\qquad \delta = \theta \beta \qquad\qquad\qquad\qquad\qquad (4.3.7)$

Once the parameters $\hat{\delta}$ and $\hat{\theta}$ are computed, the parameter vector $\beta$ can be obtained from $\hat{\beta} = \hat{\delta} / \hat{\theta}$. This dynamic efficiency model has several interesting interpretations. Assume that the reference unit $DMU_k$ is *dynamically efficient* in the sense that the following holds at t:

$$\hat{\gamma} \Delta X_{kt} + \hat{\theta}(\hat{\beta}' X_{k,t-1} - y_{k,t-1}) = \Delta y_k(t) \qquad\qquad (4.3.8)$$

Then if $\hat{\theta}$ is zero, one would have a production frontier with incremental inputs and outputs, i.e.,

$$\hat{\gamma}' \Delta X_{kt} = \Delta y_k(t) \ .$$

If one of the inputs is capital and its coefficient is $\hat{\gamma}_1$, then $\hat{\gamma}_1$ would denote the incremental output-capital ratio. This may be viewed as the short run effect. Secondly, if $\hat{\gamma}$ is a null vector and $\hat{\theta}$ is nonzero, then

$$\hat{\beta}' X_{k,t-1} = y_{k,t-1}; \hat{\beta} \geq 0 \qquad\qquad (4.3.9)$$

If the efficiency condition (4.3.8) holds for all $t \in I_T$ then (4.3.9) would indicate a steady state or a long run production frontier, where the subscript t-1 may be dropped, i.e.,

$$\hat{\beta}' X_k = y_k ; \hat{\beta} \geq 0 .$$

The parameter $\hat{\theta}$ lying between zero and one may be interpreted as the utilization coefficient. Thus we have three sources of dynamic efficiency: the level of past inputs, the change in inputs and the degree of capacity utilization. Furthermore this model (4.3.6) may easily characterize the situation when the $DMU_k$ is efficient for some t, but inefficient for others. Those DMUs, which remain dynamically efficient for all $t \in I_T$ may thus be given a special place in the peer group comparison. Finally, the parameter $\theta$ acts as a screening device between the dynamic (short run) and the static (long run) efficiency. This shows that if the utilization rates are not uniform across DMUs, the usual methods of comparing static efficiency by DEA models may be seriously biased. Now we consider an empirical application based on a cost frontier model that has been frequently used in the econometric approach to estimating the cost efficiency of a set of firms. The second application is based on the production frontier model, where dynamic changes occur over time in the various parameters.

In the first empirical application we consider the estimation of a cost frontier instead of a production frontier. By the duality theorem one can apply the cost minimization model

$$\min c(x) = \sum_{i=1}^{m} q_i x_i$$
$$\text{s.t.} \quad y < f(x_1, x_2, \ldots, x_m)$$

to derive a cost frontier, where $f(x)$ is the production function. Thus if $f(x)$ is loglinear, i.e., $\ln y = a_0 + \sum_{i=1}^{m} a_i \ln x_i$, then the dual cost frontier is loglinear also, i.e.,

$$\ln c^* = b_0 + \sum_{i=1}^{m} b_i \ln q_i + (1/r) \ln y^* \qquad (4.3.10)$$

where

$$r = \sum_{i=1}^{m} a_i, q_i = \text{input prices}, b_i = a_i/r$$

and the asterisk denotes the optimal levels. Denoting observed cost ($\ln c_j$) by $z_j$ and the explanatory variables $\ln q_i$ by $x_{ij}$ one can set up the LP model for testing the efficiency of the reference unit $DMU_k$ as follows:

$$\min h_k = \sum_{i=0}^{m+1} b_i x_{ik} = b'X_k \qquad \qquad (4.3.11)$$

$$\text{s.t.} \qquad z_j \geq b'X_j, b \geq 0; j \in I_N$$

Here the intercept term $b_0$ is subsumed in terms of the dummy input $x_{0j} = 1$ for all $j$. The empirical data used for the DEA model (4.3.11) to test the cost efficiency of unit $k$ are taken from Greene (1990).

The data set from Greene (1990) includes 123 firms in the U.S. electric utility industry, comprising the total input costs, three input prices (e.g., of capital, labor and fuel) and total output. For simplicity all input prices and input costs are expressed in relative terms by using the fuel price as a normalization factor. Thus we have in logarithmic units $x_1$ = output, $x_2$ = price of capital, $x_3$ = price of labor, $x_0 = 1$ (the dummy input) and $m = 3$. Clearly the unit $k$ is efficient, i.e., it is on the cost frontier if it satisfies for the optimal solution vector $b^* = b^*(k)$ the conditions: $z_k = z_k^* = b^{*'} X_k$ and $s_k^* = z_k - z_k^* = 0$ where $s_k^*$ is the optimal value of the slack variable. For an inefficient $DMU_k$ the observed cost $z_k$ is higher than the optimal (minimal) cost $z_k^*$, i.e., $z_k > z_k^*$, where $z_k^* = b^{*'} X_k$. By varying the objective function over $k \in I_N$ in the LP model (4.3.11) we generate two subsets $S_1$, $S_2$ of efficient and inefficient units containing $N_1$ and $N_2$ samples where $N = N_1 + N_2$.

To analyze the probability distribution of optimal costs $z_k^*, k \in S_1$ we follow the four steps outlined before. In the first step we apply the method of moments to identify the probability density function $p(z^*)$ from the set of Pearsonian curves. We restrict ourselves to the Pearson system of density functions or curves for convenience, but as Johnson and Kotz (1970) have shown that this system includes most of the frequency curves arising in practice. The identification of the density function is based on the kappa criterion, which is based on the first four moments around the man (i.e., mean, $\mu_2$, $\mu_3$, $\mu_4$) as follows:

$$\beta_1 = (\mu_3)^2 / (\mu_2)^3, \beta_2 = \mu_4 / (\mu_2)^2$$
$$k_1 = 2\beta_1 - 3\beta_1 - 6, k_2 = \beta_1 (\beta_2 + 3)^2 / [4(4\beta - 3\beta_1)(2\beta_2 - 3\beta_1 - 6)]$$
$$p(\varepsilon) = \{\beta_1 (8\beta_2 - 9\beta_1 - 12) / (4\beta_2 - 3\beta_1)\}$$
$$\qquad - \{(10\beta_2 - 12\beta_1 - 18)^2 / (\beta_2 + 3)^2\}$$

The value of $k_2$ and its sign determines which of the twelve curves fit the efficiency values $\varepsilon_j = z_j - z_j^*$ corresponding to the optimal costs. Thus if $k_2 = 0$, $\beta_1 = 0$ and $\beta_2 = 3$ one obtains the normal density for $p(\varepsilon)$. In case $k_2$ is negative one obtains the beta density. The estimated values of $\beta_1$, $\beta_2$ in our case turned out to be as follows:

$$\text{mean} = 0.170, \mu_2 = 0.021, \mu_3 = 0.004, \mu_4 = 0.002$$
$$\beta_1 = 1.973, \beta_2 = 5.105, k_1 = -1.708, k_2 = -1.308$$

This yields the beta density as follows:

$$p(\varepsilon) = 138.80 \, (1 + 6.289 \, \varepsilon)^{-0.074} \, (1—0.884 \, \varepsilon)^{5.564}$$

which defines an inverted J-shaped curve much like the exponential density. Johnson and Kotz (1970) have shown how the initial estimates of the parameters of the above beta density function p(e) can be improved upon by applying the ML method based on the method of scoring. By using this procedure the final estimate of the efficiency distribution appears as follows:

$$p(\varepsilon) = 131.21 \, (1 + 6.104 \, \varepsilon)^{-0.132} \, (1—0.723 \, \varepsilon)^{4.158}, \varepsilon \geq 0 \qquad (4.3.12)$$

This empirical density function is now used in Step 2 in the linear model

$$z_j = b'X_j + \varepsilon_j, \qquad \varepsilon_j \geq 0 \qquad (4.3.13)$$

to estimate the parameter vector b by employing the ML method, which obviously leads to nonlinear estimating equations. Two methods may be adopted by way of approximation. One is to approximate the beta density above by an exponential density, which is very close. One obtains then the form

$$p(\varepsilon) = 5.642 \, \exp(-5.642 \, \varepsilon) .$$

This can be used to derive an estimate of vector b in (4.3.13) by following the two-step method of the concentrated likelihood function applied by Schmidt (1976) in stochastic production frontier estimation. A second method is to use a truncated normal distribution, e.g., a half-normal density with range zero to infinity as an approximation and then apply the nonlinear ML method to estimate the vector b. This procedure has been adopted by Greene in stochastic production frontier estimation. Our approach differs from both these procedures in two ways. One is that the underlying efficiency distribution of $\varepsilon$ is derived from the DEA model. Secondly, the mean of the empirical efficiency distribution can be used to develop an alternative estimate by the COLS method as discussed in (4.3.3). The computed results of these different estimates are illustrated as follows in terms of the estimates of the two most important parameters of the cost and production frontier as defined in (4.3.10).

| Estimated Parameter | Beta Density (nonlinear ML) | Exponential | Half-Normal | COLS |
|---|---|---|---|---|
| Cost elasticity ($b_3$) | 0.251 | 0.210 | 0.204 | 0.215 |
| Returns to scale (r) | 3.984 | 4.762 | 4.902 | 4.651 |

Clearly the estimate based on beta density using nonlinear ML equations yields the smallest value of the returns to scale.

Finally, we compare the two efficiency distributions of cost $F_1(z_j)$ and $F_2(z_j)$ according as $z_j$ belongs to $S_1$ or $S_2$. The empirical distributions appear as follows:

| Class midpoint | $F_1(z)$ | $F_2(z)$ | Remark |
|---|---|---|---|
| >0.072 | 0.359 | 0.341 | $F_2$ dominates |
| 0.108 | 0.600 | 0.582 | $F_1$ in FSD |
| 0.179 | 0.758 | 0.604 | Sense |
| 0.250 | 0.860 | 0.614 | |
| 0.321 | 0.922 | 0.703 | |
| 0.392 | 0.960 | 0.802 | |
| 0.463 | 0.981 | 0.814 | |
| 0.534 | 0.992 | 0.891 | |
| 0.605 | 0.998 | 0.901 | |
| >0.640 | 1.000 | 1.000 | |

Clearly the distribution of the inefficient units $F_2(z)$ dominates that of the efficient units denoted by $F_1(z)$. By duality this implies that the distribution of the efficient output based on $S_1$ samples has first order stochastic dominance (FSD) over the inefficient units in sample $S_2$. Hence the mean output for $S_1$ is higher than $S_2$ and the variance for $S_1$ is equal to or lower than $S_2$.

A similar comparison of $F_1(z)$ with two other distributions resulting from the exponential and the half-normal approximation described before yielded the result that the optimal output distribution has first order stochastic dominance over the two approximations.

Now we consider the second empirical application based on the panel data set of international airlines from Cooper and Gallegos (1992). Here there is one output and three inputs. Output is measured by ton-kilometers, labor by the number of employees, fuel by the dollar expenditure and capacity by ton-kilometers available. Capacity is used as a proxy for the capital input. In our empirical application all inputs and output are measured in logarithms and the production function taken in linear and quadratic forms in logarithmic variables.

The annual data set for 11 Latin American airlines and 6 U.S. airlines over the period 1981-88 is considered for evaluating the production frontier, where a selection based on output is made of 3 airlines in each of the two groups. Thus we have used 24 observations for each group. Based on the time trend polynomial regression for each input and output we found the linear fit to be the best implying that the first difference of each series makes it stationary. Table 4.3.1 provides the estimates of dynamic DEA efficiency for the two data sets, the systematic part and the overall set. The statistical distribution of the slack variable for the LP models is summarized in Table 4.3.2. Three interesting aspects of the efficiency distribution are clearly brought out in these two Tables. First of all, the systematic efficiency provides a better fit when measured by the average residual and the ρ values. The estimate of returns to scale is much lower for the systematic components of the data set. Moreover, the proxy variable for the capital input (i.e., capacity) plays a more dominant role for the Latin airlines. Secondly, the distribution of the slack variable or the residual displays more volatility for the Latin airlines than their American counterparts. When we computed the regression of the slack variable to the capital input, the regression coefficients turn out to be 0.931 and 0.453 for the Latin group and the American group, where the first coefficient was highly significant at 1% level of the t-test. Finally, the quadratic specification provides uniformly a better fit than the linear, suggesting that the form of specification is very important in the characterization of DEA efficiency.

As we mentioned before the dynamic efficiency in DEA models can be analyzed in several ways. Four major types of dynamic efficiency are most important in a DEA framework as follows:

(a)     shift in the production or cost frontier over time,
(b)     changes in productivity parameters over time or the learning curve effects,
(c)     impact of cumulative investment on cumulative output (e.g., technical progress function ), and
(d)     the diffusion of new technology measured by different rates of adoption by different groups of firms in the industry.

For the first two cases the production response function with one output (y) and m inputs (vector x) can be specified as:

static:      $y = \beta_0 + \sum_i \beta_i x_i = \beta_0 + \beta' x$

dynamic:  $\Delta y(t) = \Delta \beta_0(t) + \beta'(t)\Delta x(t) + \Delta \beta'(t)x(t) + \Delta \beta'(t)\Delta x(t)$

where the delta operator $\Delta$ denotes the time rate of change, e.g., $\Delta y(t) = y(t+1) - y(t)$. In economic growth literature productivity growth measured by $\Delta \beta_0(t)$ represents neutral technical change, when the inputs and output are in logarithmic terms. The learning experience tends to make the inputs, particularly labor (i.e., human capital) more productive over time. Hence the vector $\Delta \beta(t)$ is important in capturing the so-called "learning by doing" effects.

Clearly the DEA model (4.3.1) in the dynamic case as above gets transformed as follows:

$$\min g_k = \Delta\beta_0 + \sum_{i=1}^{m} \left[ \beta_i \Delta x_{ik} + \Delta\beta_i x_{ik} \right]$$

$$\text{s.t. } \Delta\beta_0 + \sum_{i=1}^{m} \left[ \beta_i \Delta x_{ij} + \Delta\beta_i x_{ij} \right] \geq \Delta y_j \quad j=1,2,\ldots,N \qquad (4.3.14)$$

$$\beta_i \geq 0, \; I=1,2,\ldots,m$$

| Latin American airlines | $\beta_0^S$ | $\beta_1^S$ | $\beta_2^S$ | $\beta_3^S$ | $\beta_4^S$ | $\beta_5^S$ | $\beta_6^S$ | Avg residual | □ |
|---|---|---|---|---|---|---|---|---|---|
| Linear | 0.124 (1.45) | 0.001 (0.000) | 0.097 (0.102) | 1.036 (1.040) | - | - | - | 1.30 (2.41) | 0.907 (0.812) |
| Quadratic | 0.041 (1.021) | 0.002 (0.012 | 0.422 (0.490) | 1.230 (1.240) | 0.041 (0.050) | 0.024 (0.026) | 0.0 (0.0) | 0.17 (1.42) | 0.989 (0.840) |
| American Airlines | | | | | | | | | |
| Linear | 0.001 (0.0) | 0.011 (0.0) | 0.002 (0.0) | 0.958 (1.032) | - | - | - | 0.250 (1.89) | 0.979 (0.710) |
| Quadratic | 0.002 (0.010) | 0.001 (0.0) | 0.023 (0.0) | 0.966 (1.241) | 0.002 (0.041) | 0.011 (0.0) | 0.0 (0.0) | 0.241 (1.023) | 0.981 (0.801) |

Table 4.3.1. Estimates of systematic DEA efficiency

Notes:

The average residual or slack is defined for each LP model by the average over the excess output $(y_j^* - y_j)$ —when the excess output is positive.

The estimate $\rho$ is defined as $[1—(\Sigma\epsilon_j/\Sigma y_j)]$ where $\epsilon_j = y_j^* - y_j$ and the asterisk denotes optimal output.

The figures in the parentheses below each parameter estimate denote corresponding estimates for the whole data set comprising both systematic and unsystematic components. The coefficients for the quadratic terms are for the cross-product of inputs only.

Superscript S denotes the systematic components of the data set used in estimation.

Denote the optimal solutions (assumed to be nondegenerate for economic meaningfulness) by $\beta^*$ and $\gamma^* = \Delta\beta^*$. If the reference unit DMU$_k$ is dynamically efficient, then it holds that

$$\Delta\beta_0^* = \sum_{i=1}^{m} \left[ \beta_i^* \Delta x_{ik} + \gamma_i x_{ik} \right] = \Delta y_k$$

Otherwise it is dynamically inefficient. Thus it is clear that static efficiency need not imply dynamic efficiency.

For the airlines data we estimated in Table 4.3.3 by the static DEA model (4.3.1), the static efficiency of five Latin American airlines for each of the eight years 1981-88. To compare with dynamic efficiency we estimated for two periods 1981-84 and 1985-88 the transformed model (4.3.14) also. The results, which are reported in Table 4.3.3, show two points very clearly. First the dynamic efficiency measures are more sensitive over time. Hence the persistence of efficiency over time can be evaluated for any reference unit. Suitable time series transformations are likely to be very helpful here. Secondly, in spite of the broad similarity of the static and dynamic measures of efficiency, the two measures have very different implications for the future prospect. Thus while Avianca performs the best, Peruvian airlines reveals the most disappointing picture. Thus by using the dynamic efficiency concept and relating it to capacity or capital, one may obtain a more complete picture of efficiency.

| | | Non-zero slack or residual | | | |
|---|---|---|---|---|---|
| | Average | Minimum | Maximum | Range | Return to scale |
| Latin Am. Airlines | | | | | |
| A. Linear | 1.30 | 0.065 | 2.544 | 2.479 | 1.134 (1.1420 |
| B. Quadratic | 0.17 | 0.002 | 0.330 | 0.328 | 1.712 (1.924) |
| American Airlines | | | | | |
| A. Linear | 0.250 | 0.061 | 0.456 | 0.395 | 0.960 (1.032) |
| B. Quadratic | 0.241 | 0.017 | 0.469 | 0.452 | 1.024 (1.141) |

Table 4.3.2. Characteristics of the distribution of the residual or the slack variables for systematic efficiency.

Note: The figures in parentheses indicate the estimate of returns to scale for the overall data containing both systematic and unsystematic components.

| A. Static Efficiency | 1981 | 1982 | 1983 | 1984 | 1985 | 1986 | 1987 |
|---|---|---|---|---|---|---|---|
| LADE (E=25%) | E | E | N | N | N | N | N |
| VARIG (E=62.5%) | E | E | E | E | E | N | N |
| CRUZ (E=75%) | N | N | E | E | E | E | E |
| AVIA (E=100%) | E | E | E | E | E | E | E |
| PERU (E=0%) | N | N | N | N | N | N | N |

| B. Dyn Efficiency | 1981-84 | 1985-88 |
|---|---|---|
| LADE | N | N |
| VARIG | E | N |
| CRUZ | N | E |
| AVIA | E | E |
| PERU | N | N |

Table 4.3.3. Static and Dynamic Efficiency of Latin American Airlines

## 4.4    Stochastic Frontier Analysis

Stochastic frontier analysis (SFA) assumes a specific form of the production frontier and a composite error structure. Thus the frontier is written as

$$y_j = f(X_j; \beta) + \varepsilon_j; \quad \varepsilon_j = v_j - u_j \tag{4.4.1}$$

where $y_j$ is the single output of firm j, $j \in I_N$, $f(\cdot)$ is the production technology, $X_j$ is a vector of m inputs and $\beta$ a vector of unknown parameters to be estimated. Here $v_j$ is statistical noise assumed to be independently and identically distributed and $u_j \geq 0$ is a nonnegative error term with a fixed distribution such as gamma or truncated normal. Here $u_j$ represents technical inefficiency and $v_j$ denotes those random effects, which cannot be controlled by the firm. Various econometric methods have been attempted in recent years for estimating the parameters $\beta$ in (4.4.1) under alternative assumptions about the distribution of the error term $\varepsilon_j$. For a recent survey see Sengupta (1995a), Cornwell and Schmidt (1995) and Hjalmarsson, Kumbhakar and Heshmati (1996).

Three most recent developments in SFA are in the following areas: (1) its extension to panel data, (2) its comparison with DEA and other competing efficiency models and (3) the application to multi-output case, when both the dependent and independent variables are suitably transformed, e.g., by the Box-Cox transformation.

In case of panel data we write the frontier model (4.4.1) at time t (t=1,2,...,T) as

$$y_{jt} = f(X_{jt}; \beta) + \varepsilon_{jt}; \varepsilon_{jt} = v_{jt} - u_{jt} \tag{4.4.2}$$

Hjalmarsson, Kumbhakar and Heshmati (1996) consider a case where the production technology is represented by a translog function, e.g.,

$$\tilde{y}_{jt} = \beta_0 + \sum_{i=1}^{m} \beta_i \tilde{x}_{ijt} + \beta_t + (1/2)\left[\sum_i \sum_k \beta_{ik} \tilde{x}_{ijt} \tilde{x}_{kjt} + \beta_{jt} t^2\right]$$
$$+ \sum_i \beta_{it} \tilde{x}_{ijt} + \varepsilon_{jt} \tag{4.4.3}$$

where $\tilde{y}_{jt}$ and $\tilde{x}_{ijt}$ are the output and the inputs measured in logarithm. The estimates of returns to scale (RTS) may then be calculated from the sum of input elasticities as:

$$RTS = \sum_{i=1}^{m} \partial \tilde{y} / \partial \tilde{x}_i = \sum_i \left(\beta_i + \sum_{i=1}^{m} \beta_{ik} \tilde{x}_k + \beta_{it} t\right)$$

The time trend t is assumed to capture here exogeneous technical change as a shift in the production frontier. Technical efficiency here is

$$u_{jt} = \exp[-\eta(t - T)] u_j \equiv \eta_{jt} u_j$$

where $\eta$ is a single unknown parameter and $u_j$ is assumed for example to be normally distributed with mean $\mu$ and variance $\sigma_u^2$ truncated at zero from below. Thus the temporal variation is not the same for all firms and it increases or decreases exponentially. Given these distributional assumptions on $u_{jt}$ and $v_{jt}$, the parameters of the above model may be estimated by using the nonlinear maximum likelihood method; see Battese and Coelli (1992) for details. In case we have other explanatory variables (say z) as determinants of technical efficiency as in Reifschneider and Stevenson (1991) one may reformulate the model as

$$u_{jt} = \sum_s \delta_s z_{sjt} + w_j t; u_{jt} \geq 0 \tag{4.4.4}$$

If one assumes $w_{jt} \sim$ i.i.d. $N(0, \sigma_v^2)$ truncated at $\sum_s \delta_s z_{sjt}$ from below, then it can be shown that $(u_{jt} | \varepsilon_{jt}) \sim N(\mu_{jt}^0, \sigma_0^2)$ truncated at zero from below, where

$$\mu_{jt}^0 = \{(\sigma_v^2 \sum_s \delta_s z_{sjt} - \sigma_u^2 \varepsilon_{jt})/(\sigma_u^2 + \sigma_v^2)$$

$$\sigma_0^2 = \{\sigma_u^2 \sigma_v^2 /(\sigma_u^2 + \sigma_v^2)\} \tag{4.4.5}$$

Thus technical efficiency can be predicted from the mean, i.e.,

$$\hat{u}_{jt} = E(u_{jt} \mid \varepsilon_{jt}) = \mu_{jt}^0 + \sigma_0 \{\phi(\mu_{jt}^0 / \sigma_0)/(1 - \Phi(\mu_{jt}^0 / \sigma_0))\} \tag{4.4.6}$$

where $\Phi(\cdot)$ is the cumulative distribution of a normal density $\phi(\cdot)$. Hjalmarsson, Kumbhakar and Heshmati (1996) applied this translog frontier model to the panel data of 15 cement plants in Colombia over the period 1968-88 and compared it with the results of two types of DEA models, e.g., DEAS models where efficiency is computed each year on the basis of all observations generated up to that year (i.e., sequential) and DEAI model (intertemporal) where the data for all years are merged into one set. First, there is a fairly high rank correlation between the three measures of technical efficiency in SFA and the two DEA models. The coefficients are for example 0.58 and 0.13. However the returns to scale vary considerably across the three models, e.g., DEA models generate a large range of optimal scale levels, while SFA model indicates constant returns to scale. Secondly, all the models generate a decreasing trend in mean efficiency scores, although the levels and rates of decrease vary substantially. Thus the results for selected years appear as follows:

|      | 1968 | 1978 | 1988 | Mean |
|------|------|------|------|------|
| DEAS | 0.89 | 0.89 | 0.85 | 0.87 |
| DEAI | 0.88 | 0.72 | 0.86 | 0.77 |
| SFA  | 0.93 | 0.90 | 0.80 | 0.89 |

Finally, the use of explanatory variables such as the z variables in (4.4.4) was found to be very important when represented by the capacity utilization. Thus technical inefficiency is found to be lower in SFA models for the plants, which are utilizing more of their capacity. Other measures of the z variables such as fuel types (gas or coal) are found to be statistically insignificant.

Similar comparisons of cost frontiers by DEA and SFA models have been reported in the literature. For example Sengupta (1995b) compared two methods of establishing cost frontiers from the data set of electric utility firms from Greene (1990). One is by the corrected ordinary least squares method and the other by the DEA model. The cost frontier function is of the form

$$c_j = \beta_0 + \beta_1 y_j + \beta_2 y_j^2 + \beta_3 p_{2j} + \beta_4 p_{3j} + \varepsilon_j; j = 1, 2, ..., N \tag{4.4.7}$$

with cost (c), output (y) and the inputs of capital ($p_2$) and labor ($p_3$) expressed in logarithms (relative to fuel prices). Three specific cases are considered.

Case A. the quadratic form of (4.4.7)
Case B. the quadratic form is dropped
Case C. the term $y^2$ is replaced by the squared deviations from the mean $(y_j - \overline{y})^2$

Note that Case B is the loglinear case where $\beta_1$ may be used to test the scale economies, whereas Case C allows one to test the impact of output fluctuations on the cost frontier. The empirical estimated are reported in Table 4.4.1. The DEA estimates are derived from the least absolute deviation (LAD) approach applied to the set of DEA efficient firms. Note that in terms of the approximate t ratios the estimated slope coefficients are all statistically significant. Since the LAD estimates are closely related to median unbiasedness, they have a feature of robustness. Secondly, the evidence of significant scale economies is present in both Case A and B. However when the estimated variance term $E(y_j - \overline{y})^2$ is used as a regressor, the scale economies tend to disappear. This may suggest the role of risk adjustment in dynamic costs. Thirdly, there exists a very close correlation in efficiency estimated by the DEA and COLS method $(\rho \geq 0.48)$.

The cost frontier is more flexible than a production frontier, since it can be easily generalized to the case of multiple outputs. Thus Pulley and Braunstein (1992) provide a generalized model for multi product cost function, which can be easily transformed for applying SFA to estimate the cost frontier. Consider the Box-Cox transformation for any variable as

$$z^{(\phi)} = (z^{\phi} - 1)/\phi \text{ for } \phi \neq 0$$
$$= \ln z \text{ for } \phi = 0$$

On using this transformation a composite specification of the indirect cost function of multi product firms may be written as

$$C^{(\phi)} = f^{(\phi)}(Y, \ell nP)$$

where $Y$ is the output vector and $P$ the vector of input prices and

$$
\begin{aligned}
f^{(\phi)} = \{&[(\alpha_0 + \Sigma\alpha_j y_j^{(\pi)} + \tfrac{1}{2}\Sigma\Sigma\alpha_j\alpha_k y_j^{(\pi)} y_k^{(\pi)} \\
&+ \Sigma\Sigma\delta_{jk} y_j^{(\pi)} \ell np_k)^{(\tau)}] \\
&\cdot \exp[\beta_0 + \Sigma\beta_k \ell np_k + \tfrac{1}{2}\Sigma\Sigma\beta_{kl} \ell np_k \ell np_\ell \\
&+ \Sigma\Sigma\mu_{jk} y_j^{(\pi)} \ell np_k]\}
\end{aligned}
\tag{4.4.8}
$$

| Coefficient | DEA[a] | | | COLS | | |
|---|---|---|---|---|---|---|
| | Case A | Case B | Case C | Case A | Case B | Case C |
| $\beta_0$ | -1.44[b] | -2.62[b] | -3.01[b] | 0.01 | 0.21 | 0.34 |
| | (-0.11) | (-0.82) | (-1.45) | | | |
| $\beta_1$ | 0.00 | 0.26** | 0.29** | -0.56** | 0.23** | 0.37** |
| | (0.0) | (25.9) | (22.1) | (-11.68) | (9.03) | (27.48) |
| $\beta_2$ | 0.02** | - | 0.04** | 0.06** | - | 0.06** |
| | (16.0) | | (16.0) | (16.73) | | (16.5) |
| $\beta_3$ | 0.09 | 0.12* | 0.15* | 0.29** | 0.41* | 0.32** |
| | (1.43) | (1.87) | (1.83) | (3.43) | (1.98) | (3.75) |
| $\beta_4$ | 0.61** | 0.59** | 0.57** | 0.38** | 0.33 | 0.36** |
| | (0.979) | (9.45) | (7.24) | (4.57) | (1.63) | (4.29) |
| $R^2$ | - | - | - | 0.939 | 0.623 | 0.938 |
| App. $R^2$ | 0.518 | 0.265 | 0.729 | - | - | - |
| Elasticity | 0.273 | 0.259 | 1.004 | 0.374 | 0.225 | 1.309 |

Table 4.4.1. Cost frontier estimates: DEA and COLS

[a]The t-ratios in parentheses for the DEA estimates are derived from the least absolute deviation (LAD) method based on bootstrap standard error; also, one and two asterisks denote t-values significant at 5% and 1% levels respectively, e.g., see Hardle and Bowman (1988) for the bootstrap algorithm.

[b]The negative values of the intercept term are due to the inequality $\beta_0 < 0$ imposed on the LP models.

App.$R^2$ is an approximate measure of $R^2$ defined as $1 - (\Sigma s_i^2 / \Sigma[c_i - med(c)]^2)$ where s is the slack variable, c = cost and med(c) is median cost.

For different restrictions on the transformation parameters $\phi$, $\pi$ and $\tau$ this generalized model nests the translog, a separable quadratic specification and a composite function. For instance the restrictions $\phi = 0$, $\pi = 0$ and $\tau = 1$ yield the translog cost function, whereas $\pi = 1$, $\tau = 0$, $\delta_{jk} = 0 = \mu_{jk}$ yield a separable quadratic specification.

Assuming the errors for model (4.4.8) to be normally distributed, the log-likelihood function for the sample observations $C_1, C_2, ..., C_T$ is given by

$$L = -(T/2)\ell n\sigma^2 - [1/2\sigma^2]\Sigma(C_t^{(\phi)} - f^{(\phi)}(Y, \ell nP))^2$$
$$+ (\phi - 1)\Sigma\ell nC_t$$

where $(\phi - 1) \ln C_t$ is the logarithm of the Jacobian of the transformation from $C_t$ to $C_t^{(\phi)}$. Conditional on the data and for fixed values of the parameters the maximum likelihood estimate (MLE) of the error variance $\sigma^2$ is given by

$$(1/T)\Sigma(C_t^{(\phi)} - f^{(\phi)}(Y, \ell nP))$$

Let $C^*$ be the geometric mean of the sample observations. On substituting $C^*$ and the MLE of $\sigma^2$ one obtains the concentrated log-likelihood function for the parameters

$$L(\phi, \Omega) = -(T/2)\ell n[(1/T)\Sigma\{(C_t^{(\phi)} - f^{(\phi)}(Y, \ell nP))/(C^*)^{\phi-1}\}^2]$$

where $\Omega$ represents all cost function parameters other than $\phi$. Hence the MLEs of the parameters of the cost function minimize the expression

$$\Sigma[(C_t^{(\phi)} - f^{(\phi)}(Y, \ell nP))/(C^*)^{\phi-1}]^2$$

which means that standard nonlinear least squares routines can be employed to estimate $\phi$ and other parameters in $\Omega$. For computing algorithms see Carroll and Ruppert (1988).

Pulley and Braunstein (1992) applied this generalized cost function model to the 1988 sample data of 205 banks with assets exceeding one billion dollars. Four output categories measured in dollars are used here, e.g., demand deposits plus savings, real estate loans, commercial loans, and credit card loans. They estimate economies of scope as the cost savings from simultaneously producing all four outputs in the same firm, rather than producing each output separately in a specialized firm. Thus the cost savings ranged from 27.07 for the general model (4.4.8) to 755.8 in the generalized translog model and 0.44 in case of the separable quadratic cost function.

Two points are to be noted. The form of Box-Cox transformation may vary depending on the specific form of distribution of the error term $\varepsilon_j$ introduced in (4.4.1) for example. But the approximate normality of errors resulting from the transformation makes it suitable for least squares type estimation. Secondly, the estimates of parameters of the model (4.4.8) vary widely depending both on the form of transformation and the algorithm used for nonlinear estimation. This emphasizes the need for robustness analysis in this SFA framework.

## 4.5    Statistical Tests of Efficiency

Statistical tests of efficiency scores in DEA model depend on the underlying stochastic data used in the DEA model. Since the statistical distribution of input output data is not directly considered in the DEA a nonparametric procedure is more appropriate here. Two types of nonparametric tests are most readily applicable here. One is based on the Kolmogorov-Smirnov (KS) statistic, which may be used to test the statistical fit of the distribution over the set $S_1$ of efficient output or cost levels as determined by the DEA model. Sengupta (1989) has applied this method to determine the cost efficiency of public electric utilities. Since the DEA model determines two subsets of DMUs, one efficient ($S_1$) and the other inefficient ($S_2$) this KS statistics may also be applied to test the statistical difference (or distance) between these two distributions. A second type of test is based on a bootstrapping technique applied recently by Simar and Wilson (1998) to the efficiency scores $\theta$ obtained for example from the technical efficiency model (1.4). Bootstrapping is based on the idea of repeatedly simulating the data generating process through resampling and then applying the estimates mimicing the sampling distribution. In this section we describe briefly these two types of applied nonparametric tests.

Consider the subset $S_1$ of efficient cost levels determined by the DEA model (4.4.11). In the nonparametric case we estimate the probability density $p_H(z)$ of $z_j$ over the subset $S_1$ that includes the units found to be efficient by the DEA method. We partition the closed interval [a,b] of z by $a = z_{(0)} < z_{(1)} < z_{(2)} < z_{(k)} = b$ and consider the histogram estimates of the form

$$p_H(z) = \begin{cases} c_r & \text{for } z_r < z < z_{(r+1)}, r = 0,1,...,k-1 \\ c_{r-1} & \text{for } z_{(k)} = b \\ 0 & \text{otherwise} \end{cases} \qquad (4.5.1)$$

where $p_H(z) > 0$ and $\int_a^b p_H(z)dz = 1$. To estimate the population histogram of this form we consider the entire sample space in subset $S_1$ and count the number of observations falling in the r-th interval. Let $N_r$ be this number. Then the population parameter $c_r$ above can estimated by $\hat{c}_r = N_r[N(z_{(r+1)} - z_{(r)})]^{-1}$ for $r = 0,1,...,k-1$.

A similar method can be used to determine the density $p_{\tilde{H}}(z)$ of costs over the subset $S_2$ of inefficient units. Let $F(z)$ and $\tilde{F}(z)$ be the corresponding cumulative distributions. Then the divergence of the two distributions may be tested by the KS statistic $D_{m_1,m_2}$

$$D_{m_1,m_2} = \max \left| \left(F_{m_1}(t) - \tilde{F}_{m_2}(t)\right)\left(\frac{m_1 m_2}{m_1 + m_2}\right)^{-1/2} \right|$$

where $F_{m_1}(t) = (1/m_1)$ (number of optimal values less than or equal to t) and likewise for $\tilde{F}_{m_2}(t)$. If the null hypothesis is rejected, then the efficiency frontier is statistically different from the inefficiency profile.

Several other statistical methods are also available for testing the statistical difference of observed data from the frontier data. First of all, one may statistically fit regression functions over the two subsets $S_1$ and $S_2$ determined by the DEA model. Then Chow-test can be applied to test their difference. This is of course a parametric test. A second method, which is nonparametric is based on the bootstrapping technique, which has been applied, recently by Lothgren and Tambour (1997). They assume that outputs are generated by random radial deviations off the isoquant of the output set. Formally the input output observations for each of the $k \in K$ firms are given by

$$(x_k, y_k) = (x_k, y_k^f / F_{0k}) \tag{4.5.2}$$

where $F_{0k}$ is a stochastic efficiency measure defined as $F_0(y, x) = \max\{\lambda : \lambda y \in P(x)\}$, where $P(x)$ is the production possibility set and $y_k^f$ is the unobservable frontier output for the kth firm. The efficiency measures for the K firms are assumed to be drawn from the same common distribution. The bootstrap method is to mimic in each resample the data generating process underlying (4.5.2). Conditioned on observed inputs and outputs, the resampled data are used to obtain bootstrap estimates of the efficiency measures and this set of measures is used to perform the hypothesis tests on scale efficiency and other tests.

Simar and Wilson's method has two additional features. First of all, the reflection and kernel smoothing methods are applied before resampling the original efficiency measures. Secondly, the resampling method proposed by Simar and Wilson emphasizes on the sampling variability of the production frontier estimate underlying the efficiency distribution, whereas Lothgren and Tambour explicitly allow for the fact that the sampling distribution of the DEA frontier is a function of both the sampling distributions of the frontier estimate and of the underlying input output data.

One advantage of the bootstrap method is that this simulation experiment can be used over cross section rather than panel data to analyze the sensitivity of nonparametric efficiency scores to sampling variations. Thus it provides a robustness test. The difficulty is that it requires a large amount of data generation in the various simulation experiments needed for this method.

We may also refer to one other nonparametric test typically applicable to panel data. This is the rank test known as the Kruskal-Wallis test, which has been recently applied for example by Sueyoshi (1999) over the time series (1983-1997) data of Japanese postal services. The efficiency scores are here ranked by

$R_j$ from the smallest ($R_j = 1$) to the highest ($R_j = N$) and then the statistic H is formed

$$H = \left[ \frac{12}{N(N+1)} \sum_{t=1}^{T} (W_p^2 / n) \right] - 3(N+1) \qquad (4.5.3)$$

where    $W_p = \Sigma R_j$ .

This rank statistic is approximately distributed as chi-square with T-1 degrees of freedom under the condition that the number (n) of DMUs in each year exceeds five. Here N is the total number of DMUs when the data set is accumulated as a whole over the entire period. Sueyoshi applied this rank test to statistically examine whether a frontier shift has occurred over some periods.

# 5. Industrial Applications

From a historical perspective the applications of nonparametric efficiency measurement by the DEA approach have passed through three stages. Initially Farrell (1957) applied this method to a cross-section of agricultural farms using only input and output data. Hence only production or technical efficiency was estimated. Although input and output price data were available, Farrell chose not to use it because such data contained random noises and fluctuations. Thus allocative or price efficiency was not examined in any detail, although he mentioned that it is likely to be very useful in comparing various industries rather than firms. Farrell's method was however restricted to one output. The extension to many outputs and the engineering concept of a ratio of weighted outputs to weighted inputs were pioneered by Charnes, Cooper and Rhodes (1978). Charnes, Cooper and his associates have made series of applications. Charnes et al (1994) has recently surveyed the theory and applications of this field and Sengupta (1995a) has considered some dynamic and stochastic extensions. Most of the applications in this phase are for public sector organizations like schools, clinics and hospitals, army training units and prison systems. More recently private sector applications to commercial banks, international airlines, chains of groceries and restaurants and telephone operations are increasingly being made. These applications bring forth several new issues and challenges for the DEA approach to efficiency measurement.

For example the profit measure has to be compared with the technical and the allocative efficiencies derived by the DEA approach and frequently these two measures do not go in the same direction. Secondly, the market structure of demand plays a significant role on the pattern of output supply by firms. The economic theory of monopolistic competition with the influence of advertising expenditures on total revenue and costs becomes very important in this framework. Finally, when different firms are in oligopolistic competition, the concept of efficiency must be viewed in terms of a game-theoretic equilibrium and in this setup the traditional definition of DEA efficiency no longer holds.

## 5.1    DEA Models for Competition

Economic theory predicts that the competitive pressure in market demand tends to improve cost efficiency of individual firms in an industry.

Our objective in this section is to develop a class of efficiency models in the semi and nonparametric framework, where market competition and price uncertainty influence the technical or production efficiency of a set of firms comprising the industry. Farrell (1957), who is the first to develop the nonparametric method of efficiency measurement recommended against the use of prices in determining allocative efficiency mainly because of the fluctuations of demand and market prices. Two motivations may be given for considering market demand and its influence on economic efficiency viewed as allocative efficiency, which measures the firm's success in choosing an optimal set of inputs with a given set of input and output prices or costs. One is the case of imperfect competition, where firms have some market power though limited in influencing the market price of output. The second motivation is that the measures of profitability are not adequate in themselves for comparing private sector units or firms on their performance. This is because firms operate in different environments, which affect profit and these environmental factors are outside the control of units and must be allowed for if an appropriate and fair assessment of their performance is to be made. Thus Schefczyk (1993) found for international airlines data of 14 airlines over the period 1988-90 that gross profit margins do not explain more than 32% of their technical or production efficiency, which measures the firm's success in producing maximum output from a given set of inputs. A similar result was found by Athanassopoulos and Thanassoulis (1995), who argued for separating market or price efficiency from profitability, and applied the DEA model to assess market efficiency evaluation. The first stage relates to the firm attracting demand and generating revenue. This stage is called *'market efficiency'*. Advertising expenditure and marketing cost are utilized in this stage to optimally influence the market demand. The second stage minimizes operating costs for a given revenue level and it is termed *'cost efficiency'*. These two stages thus provide a basis for generating optimal profits. In the standard nonparametric approach to production or technical efficiency adopted by Farrell (1957) and the DEA researchers the market efficiency aspect is completely ignored. Implicitly this assumes either that demand is already predetermined, or that each firm is a price taker selling as much as it can at the lowest possible operating cost. Under conditions of imperfect competition when market shares have some influence on the equilibrium output price, such an assumption may not be very appropriate. Hence we may consider a two-stage approach while applying the nonparametric method to efficiency measurement to firms or enterprises in a 'large group' sense in a Chamberlinian framework.

## A. Allocative Models of Efficiency

We consider first a model of relative efficiency, where a firm or decision-making unit (DMU) is compared to the cluster of firms or DMUs and only input output data are available with no price data. This is the comparison of production or technical efficiency across firms. Secondly, we consider a model, which minimizes overall unit costs subject to input and output constraints. This yields overall efficiency, which can be decomposed into technical and allocative efficiency. Finally, we consider a more general class of models where demand considerations are introduced and both demand and cost uncertainty are incorporated. This type of model is generalized to dynamic frameworks, when intertemporal cost functions are minimized. Consider first a DEA model for characterizing the static efficiency of a reference unit k in a cluster of N units, where each unit j or $DMU_j$ has m inputs ($x_{ij}$) and s outputs ($y_{rj}$):

$$\underset{\lambda,\theta}{\text{Min}\,\theta}$$

$$\text{s.t.}\quad \sum_{j=1}^{N} x_{ij}\lambda_j \leq \theta x_{ik}; \sum_{j=1}^{N} y_{rj}\lambda_j \geq y_{rk} \tag{5.1.1}$$

$$\sum_{j=1}^{N} \lambda_j = 1, \lambda_j \geq 0, \theta \geq 0$$

In vector matrix form this is:

$$\text{Min}\,\theta, \text{s.t.} X\lambda \leq \theta X_k; Y\lambda \geq Y_k; \lambda'e = 1; \lambda \geq 0 \tag{5.1.2}$$

where e is a column vector with N elements, each of which is unity and the prime denotes a transpose. Here the input ($X_j$) and output ($Y_j$) vectors (j=1,2,...,N) are all observed and this is called an input oriented model in DEA literature. Here the reference unit k is compared with the other (N-1) units in the cluster. Let $\lambda^* = (\lambda_j^*)$ and $\theta^*$ be the optimal solutions of the above DEA model with all the slack variables zero. Then the reference unit k or $DMU_k$ is technically efficient if $\theta^* = 1$ and the first two sets of inequalities in (5.1.1) hold with equality. Thus the optimal value of $\theta^*$ provides a measure of technical efficiency (TE). If $\theta^*$ is positive but less than unity, then it is not technically efficient at the 100 percent level. Overall efficiency ($OE_j$) of a DMU or firm j however combines both technical ($TE_j$) or production efficiency and the allocative ($AE_j$) or price efficiency as follows:

$$OE_j = TE_j \times AE_j; j = 1, 2, ..., N \tag{5.1.3}$$

To characterize overall efficiency of the reference unit $DMU_k$ one sets up the linear programming (LP) model as follows:

$$\underset{x,\lambda}{\text{Min}\,q'x}$$

$$\text{s.t.}\quad \sum_{j=1}^{N} X_j\lambda_j \leq x; \sum_{j=1}^{N} Y_j\lambda_j \geq Y_k \tag{5.1.4}$$

$$\lambda'e = 1; \lambda \geq 0; x \geq 0$$

Here $q$ is an m-element vector of unit costs or input prices as observed in the competitive market and $x$ is an input vector to be optimally decided by $DMU_k$ along with the weights $\lambda_j$. Here $X_k$ and $Y_k$ are the observed input and output vectors for the reference unit $k$, whereas $x$ is the unknown decision vector to be optimally determined. Let $\lambda^*$ and $x^*$ be the optimal solution of the LP model (5.1.3) with all slacks zero. Then the minimum input cost is given by $c_k^* = q'x^*$, whereas the observed cost of the reference unit is $c_k = q'X_k$. Hence the three efficiency measures are defined as follows:

$$TE_k = \theta^*; OE_k = c_k^*/c_k \text{ and } AE_k = OE_k/TE_k \qquad (5.1.5)$$

Note that the input vector $x$ in (5.1.4) is a decision vector to be optimally chosen, whereas $X_k$ is the observed data in (5.1.1). If $\theta^*X_k = x^*$, then the two models generate identical optimal solutions; otherwise the two optimal solutions are very different. The dual problems corresponding to (5.1.4) and (5.1.1) appear as follows:

$$\text{Max } \alpha'Y_k + \alpha_0$$
$$\text{s.t.} \quad \beta \leq q \text{ and } \beta'X_j \geq \alpha Y_j + \alpha_0; j = 1, 2, ..., N \qquad (5.1.6)$$
$$\alpha, \beta \geq 0, \alpha_0 \text{ free in sign}$$

and

$$\text{Max } \alpha Y_k + \alpha_0$$
$$\text{s.t.} \quad \beta'X_j \geq \alpha Y_j + \alpha_0; j = 1, 2, ..., N \qquad (5.1.7)$$
$$\alpha, \beta \geq 0, \alpha_0 \text{ free in sign.}$$

Let asterisks denote optimal values and let $DMU_k$ be efficient. Then it must follow from (5.1.6) that the production frontier for the k-th unit is as follows:

$$\alpha^{*'} Y_k = \beta^{*'} X_k - \alpha_0^*$$

but since $\beta^*$ is constrained as $\beta^* \leq q$, we must have $\alpha^{*'} Y_k \leq q' X_k - \alpha_0$. Thus so long as the actual inputs $X_k$ are not equal to their optimal levels $x^*$, this efficiency gap measured by $(q'X_k - \alpha_0^* - \alpha^{*'} Y_k)$ may persist. Thus the constraint $\beta^* \leq q$ reflects the fact that the observed input $X_k$ of the reference unit may or may not be equal to the optimal level $x^*$, when all firms face the same competitive price $q$. There is no such constraint for the dual problem (5.1.7).

Note that if $\alpha_0^*$ is positive (negative or zero), then we have increasing (decreasing or constant) returns to scale.

We now consider a more generalized version of the overall efficiency model (5.1.4) where both input (q) and output prices (p) are assumed to be available and the optimal vectors on input and output are optimally chosen as follows:

$$\text{Max } p'y - q'x$$
$$\text{s.t.} \quad \sum_{j=1}^{N} X_j \lambda_j \leq x; \sum_{j=1}^{N} Y_j \lambda_j \geq y \qquad (5.1.8)$$
$$x \leq X_k; y \geq Y_k; \lambda'e = 1 \qquad x, y, \lambda \geq 0$$

Here (x,y) are the control vectors of inputs and output to be optimally determined and $(X_k, Y_k)$ denote the observed levels for $DMU_k$. The dual of this problem then becomes

$$\text{Max } v'Y_k - u'X_k - \alpha_0$$
$$\text{s.t.} \quad p \leq \alpha - v, q \geq \beta - u \qquad (5.1.9)$$
$$\beta'X_j \geq \alpha'Y_j + \alpha_0, j = 1, 2, ..., N$$
$$(u,v,\alpha,\beta) \geq 0, \alpha_0 \text{ free in sign}$$

Two special cases of the generalized model (5.1.8) are of great importance. One is the simpler output-oriented model where demand $(d_r)$ for output $(y_r)$ is subject to a probability distribution $F(d_r)$ and the objective function is to maximize the expected value of total revenue minus expected inventory cost. This yields the model

$$\text{Max } E\left[ \sum_{r=1}^{s} p_r \min(y_r, d_r) - \sum_{r=1}^{s} h_r(y_r - d_r) \right] \qquad (5.1.10)$$
$$\text{s.t.} \quad X\lambda \leq X_k; X\lambda \geq y, \lambda'e = 1, \lambda \geq 0$$

where $\lambda$ and y are the unknown vectors to be optimally solved for and $h_r$ is the observed unit cost of positive inventory for $y_r > d_r$. Denoting optimal values by asterisks, the efficient $DMU_k$ would then satisfy the following marginal condition:

$$F(y_r^*) = (p_r + h_r)^{-1}(p_r - \alpha_r^*)$$
$$\alpha^{*\prime} Y_k = \beta^{*\prime} X_k - \alpha_0^*$$

Clearly higher output price and lower inventory costs would increase the optimal output levels $y_r^*$, which may be compared with the observed output $y_{rk}$ in output vector $Y_k$.

The second case is an input-oriented model, where the input decision $x_i$ is equal to planned values $\bar{x}_i$ plus an error term $\varepsilon_i$ with a zero mean and fixed variance. The errors are disturbances such as mistakes or unexpected difficulties in implementing a planned value $\bar{x}_i$. The planned values $\bar{x}_i$ are the decision variables, which have to be optimally chosen by each DMU and the error process $\varepsilon_I$ is realized after the planned value of $x_i$ is optimally selected. The input constraints now turn out to be chance constrained.

$$\Pr ob\left[ \sum_{j=1}^{N} x_{ij}\lambda_{ij} \leq \bar{x}_i + \varepsilon_i \right] = \delta_i, 0 < \delta_i < 1$$

where $\delta_I$ is the tolerance level of the ith input constraint. The simpler model then takes the following form:

$$\text{Min} \sum_{i=1}^{m} q_i \bar{x}_i$$

$$\text{s.t.} \quad \sum_{j=1}^{N} x_{ij}\lambda_j = \bar{x}_i + w_i; w_i = F^{-1}(1 - \delta_i) \qquad (5.1.11)$$

$$\sum_{j=1}^{N} y_{rj}\lambda_j \geq y_{rk}; \lambda'e = 1, \lambda \geq 0 \quad i=1,2,\ldots,m; r=1,2,\ldots,s$$

Clearly the input uncertainty is here captured by the term $w_i$ which depends on the level $\delta_I$ of chance constraint, e.g., the higher the level of $w_i$, the lower would be the optimal planned inputs $\bar{x}_i^*$.

We assume now the role of market competition in the efficiency framework. Hence we assume that each firm or $DMU_j$ produces a single homogeneous output denoted by $y_j$, where the total industry output is denoted by $y_T = \sum_{j=1}^{N} y_j$. If N is large and the firms or DMUs are competitive, then the output price p is a constant, unaffected by the size of each individual firm. In this case the price can be viewed as $p = \bar{p} + \varepsilon$ made up of two components: the expected price $\bar{p}$ and a random part $\varepsilon$ with a zero mean and a constant variance $\sigma_e^2$. The total cost of inputs for each firm may now be related to output as

$$c(y_j) = c\,y_j + F_j$$

assuming a linear form, where $F_j$ is the fixed cost and c is marginal cost that is assumed to be identical for each firm. Maximization of expected profits would then yield the LP model:

$$\text{Max }\ \bar{\pi} = (\bar{p} - c)y - F$$

$$\text{s.t.}\quad \sum_{j=1}^{N} X_j \lambda_j \le X_k;\ \sum_{j=1}^{N} y_j \lambda_j \ge y;\ \lambda' e = 1, \lambda \ge 0 \qquad (5.1.12)$$

where y is the unknown decision variable to be optimally selected. In case the market is imperfectly competitive, the price variable then depends on the output supply of different firms. In the homogeneous output case the firms are all alike, and the inverted demand function can be written as:

$$\bar{p} = a - bY_T, Y_T = \sum_{j=1}^{N} y_j; y_k = y$$

The LP model (5.1.12) would then yield the following optimality conditions:

$$(a - c) - bY_T - by* - \alpha* \le 0$$

$$\alpha* y_j - \beta*' X_j - \alpha_0^* \le 0 \qquad (5.1.13)$$

$$\alpha*, \beta* \ge 0, \alpha_0^* \text{ free in sign}$$

If firm k is efficient, then one must have

$$y* = (a_1 / b) - Y_T - \frac{\alpha*}{b}; y* > 0$$

$$a_1 = a - c > 0$$

where $y* = y_k^*$ is the efficient output of the k-th firm. If all firms are efficient, then $Y_T^* = \sum_{j=1}^{N} y_j^*$ and one obtains

$$Y_T^* = (N/b)(1 + N)^{-1}[a_1 - \alpha*] \qquad (5.1.14)$$

Now we introduce organizational slack denoted by $s_k$ in the cost function

$$c(y) = (c + s_k) y + F_k; s_k > 0$$

Recently Selten (1986) has interpreted this slack concept due to Leibenstein's (1966) X-efficiency as a part of the cost function and introduced a 'strong-slack' hypothesis which maintains that this type of slack has a tendency to increase so long as profits are positive, i.e., this slack can be reduced only under the threat of losses. Including this slack-ridden cost into the profit maximization model would yield the optimality condition for the efficient output as:

$$y^* = (a_1 - \alpha - s_k)/b - Y_T; \ y^* > 0$$

where $y^* = y_k^*$ is the efficient output of the firm k. If all firms are efficient then

$$Y_T^* = (N/b)(1+N)^{-1}[a_1 - \overline{s} - \alpha^*]$$
$$\overline{p} = a + N(N+1)^{-1}(\alpha^* + \overline{s} - a_1) \qquad (5.1.15)$$
$$\pi_k^* = (\overline{p} - c\, s_k)y^* = F_k$$

when $\overline{s} = \sum\limits_{k=1}^{N} s_k/N$ is the average rate of slack. Several implications follow from this set (5.1.15) of efficiency conditions. First of all, the long run pressure of competition would tend to lead to zero profits $\pi_k^* = 0$ for all k=1,2,...,N according to the strong slack hypothesis. In this case the expected price becomes $\overline{p} = c + \overline{s} + (F/y^*)$. This shows that fixed costs have a strong positive role in determining the long run equilibrium price. The higher the average slack rate $\overline{s}$, the higher is the equilibrium expected price. Secondly, as the number of N firms increases, it increases the volume of total industry output $Y_T^*$ and reduces the average price. Finally, as the average slack rate $\overline{s}$ rises (falls), it increases (decreases) the equilibrium price. Note that in case of weak slack hypothesis all profits are not squeezed out and there remains a divergence of individual ($s_k$) from the average slack rate ($\overline{s}$), when the latter is positive. Thus some inefficiency may persist due to the existence of a positive slack.

So far we have assumed that the expected price $\overline{p}$ is the market-clearing price equating market demand and supply. If however this is not the case, then the supply y would differ from demand d, where demand is subject to random fluctuations around the mean level $\overline{d}$. In this framework we have to add to the cost function the cost of inventory and storage C(y-d). Assuming this cost to be quadratic one may then formalize the decision model

$$\text{Max } \overline{\pi} = (\overline{p} - c - s_k)y - (1/2)\gamma E(y-d)^2 - F \qquad (5.1.16)$$
$$\text{s.t.} \quad \text{the same constraint as in (5.1.12)}$$

In this case the optimality conditions for the efficient output becomes

$$y^* = (b + \gamma)^{-1}[(a_1 - \alpha^* - s_k) + \gamma\bar{d} - bY_T] \tag{5.1.17}$$

where $\gamma$ is the unit cost of inventory or shortage and $\bar{d}$ is the expected level of demand. In this case the marginal impact $\partial y^*/\partial\gamma$ of inventory/excess costs may be either negative or positive according as

$$b(Y_T + \bar{d}) > \text{or}, < (a_1 - \alpha^* - s_k)$$

Again this explains the persistence of some inefficiency, when demand is uncertain and the firm chooses its optimal output by the quadratic criterion of adjusted profits. Furthermore, the higher the mean demand the greater the optimal level of efficient output $y^*$. In case of perfect competition with each firm a price taker, the optimality condition (5.1.17) reduces to

$$y^* = \bar{d} + (1/\gamma)(\bar{p} - c - s_k - \alpha^*) \tag{5.1.18}$$

which shows unequivocally that higher inventory costs ($\gamma$) lead to lower optimal output. Again by comparing the observed output $y_k$ with the optimal output $y^*$, one could evaluate the impact of inefficiency. Note that we still have the comparative static results: $\partial y^*/\partial\bar{p} > 0$ and $\partial y^*/\partial s_k < 0$. Since $\bar{p} = \gamma(y^* - \bar{d}) + c + s_k + \alpha^*$, we have the results:

$$\bar{p} > MC_T, \text{ if } y^* > \bar{d}$$

and

$$\bar{p} < MC_T, \text{ if } y^* < \bar{d} \tag{5.1.19}$$

where $MC_T = c + s_k + \alpha^*$ is total marginal cost with three components: production costs (c), cost of slack ($s_k$) and the cost of discrepancy of observed from optimal output ($\alpha^*$) Clearly the case of multiple outputs can be handled in a symmetrical way.

## B. Market and Cost Efficiency: Two-Stage Model

The two-stage efficiency model solves for optimal revenue, net of advertising costs in the first stage, where the market demand characteristics are directly incorporated into the DEA model. Given this optimal net revenue the second stage minimizes the total operating costs so as to determine the cost efficient unit or firm. Clearly under conditions of market uncertainty and imperfect competition, a firm may not be fully market efficient in the first stage, although it may turn out to be cost efficient in terms of the DEA model in the second stage. The second stage may also be given a dynamic interpretation over time in terms of the adjustment cost theory. Fr example the firm may add an adjustment cost to the overall objective of minimizing operating costs, where the adjustment cost may reflect the costs of deviation from the optimal revenue and its implied input requirements determined in the first stage. Recently Callen, Hall and Henry (1990) and Sengupta (1994) have used this adjustment cost approach in a quadratic form to capture the producer's risk averse attitude towards input and output fluctuations.

Consider first the market efficiency model in a DEA framework, where we assume for simplicity one output but m advertisement inputs ($A_i$) with unit costs $c_i$. The firm minimizes net revenue:

$$\text{Max NR} = pf_k My - \sum_{i=1}^{M} c_i A_i$$

$$\text{s.t.} \quad \sum_{j=1}^{N} A_{ij}\mu_j \le A_i; \sum_{j=1}^{N} y_j\mu_j \ge y \qquad\qquad (5.1.20)$$

$$\Sigma\mu_j = 1, \mu_j \ge 0; \ j=1,2,\ldots,N$$

by choosing the optimal levels of output (demand) y and the advertisement inputs of different types. Here p is the output price which depends on both demand y facing the firm and its advertisement costs, i.e., $R = py = R(y, A_1,\ldots,A_2)$ where R is gross revenue. This type of model is most suitable for stores under a grocery chain, restaurants or branches of different banks, where $f_k$ is the fraction of customers out of the potential total M of customers who buy from the store k, whose optimal levels of advertisement $A_i$ and demand y are to be determined. Let $A^* = (A_i^*)$ and $y^*$ be the optimal solutions of the LP model (5.1.20) with a given M. then the unit or firm k has market efficiency at 100% level if the following conditions hold:

$$\frac{Mf_k R^*}{A_i^*}\left(\varepsilon_{R \cdot A_i} + f_f \cdot A_i\right) + \beta_i^* - c_i = 0$$

$$\text{for } i=1,2,\ldots,m \qquad\qquad (5.1.21)$$

$$p^* f_k M\left(1 - \frac{1}{\varepsilon_{y \cdot p}}\right) = \alpha^*$$

$$A_k'\beta^* = \alpha^* y^* - \alpha_0^*; ; (\alpha,\beta) \ge 0; \ \alpha_0 \text{ free in sign}$$

where the Lagrangean expression is

$$L = p(y, A)f_k My - c'A + \beta'(A - \sum_j A_j \mu_j) + \alpha(\Sigma y_j \mu_j - y) + \alpha_0(1 - \Sigma \mu_j)$$

and the term $\varepsilon_{z \cdot x}$ denotes the elasticity of z with respect to x, e.g., $\varepsilon_{f_k \cdot A_i}$ denotes the elasticity of the market share $f_k$ with respect to the advertisement expenditure $A_i$, i.e., $\varepsilon_{f_k \cdot A_i} = \dfrac{\partial f_k}{\partial A_i} \cdot \dfrac{A_i}{f_k}$. Note that the market efficiency concept here involves three components, e.g., (1) the optimal levels of advertisement expenditures (or marketing inputs), which incorporate the response of market share and market revenue, (2) the optimal levels of demand or output y* of the k-th firm, so that if the observed output $y_k$ is not equal to y* then we have market inefficiency and finally (3) an optimal marketing response frontier, where the optimal firm demand depends on the shadow prices of the marketing input-mix. Thus the market inefficiency may be decomposed into these three levels. Note that in perfect competition one can drop out all the elasticity terms and then obtain for the efficient unit k:

$$\beta_i^* = c_i; p * f_k M = \alpha^*; A_k' \beta^* = \alpha^* y^* - \alpha_0^*$$

Clearly this is the condition that price equals marginal cost.

Note that the first stage decision model allows the firm to expand the market to an optimal size. This is a very important consideration in the world market today, where the openness in trade and liberalization have expanded the world export market and some firms have achieved phenomenal success in achieving strong economies of scale, which can keep production costs down and also increase the productivity of inputs.

This implies that the firms, which achieve market efficiency may profitably exploit methods of reducing total operating costs (TOC) in the short run and the capacity costs in the long run. For the short run cost minimization problem one may extend the DEA model (5.1.14) by adjoining a vector of research inputs ($R_j$) for each firm j:

Min TOC $= q'x + C(R)$

$$\text{s.t.} \quad \sum_{j=1}^{N} x_j \lambda_j \le x; \sum_j R_j \lambda_j \le R \qquad (5.1.22)$$

$$\sum_j Y_j \lambda_j \le Y_k; \lambda'e = 1, \lambda \ge 0$$

where it is assumed that maximum revenue for firm k is predetermined in the first stage. Here C(R) denotes the total cost of research inputs. If the research cost is linear $C(R) = \Sigma g_i R_i$, then the LP model (5.1.22) may be used for determining the efficient levels of inputs $x_i^*$ and $R_w^*$ and the efficient level of the output vector $y^*$. Then firm k's cost inefficiency may be measured in terms of the gaps: $X_k \neq *, R_k \neq R^*$ and $y^* \neq d$. But the research inputs tend to lower the input costs of production. To capture this aspect one may assume an input cost function, where the R&D inputs tend to lower the initial unit production cost $q_i$, i.e.,

$$TOC = \sum_{i=1}^{m} (q_i - f_i)x_i + \sum_{w=1}^{W} h_w R_w$$

Here $f_i$ is the marginal reduction of unit production cost $q_i$, and $h_w$ is the observed cost of the research input $R_w$. This type of cost-reducing process innovation through R&D has been used by d'Aspremont and Jacquemin (1988) in a quadratic form in a Cournot-type cooperative R&D model. In case of the long run the output vector $y_t$ has to be constrained at each time point by the capacity vector $\overline{y}$ where the cost of capacity $C(\overline{y})$ has to be adjoined to the objective function. This yields the dynamic TOC minimization model

$$\text{Min TOC} = \sum_{t=1}^{T} [q_t' x_t + C(R_t) + C(\overline{y}_t) + C(y_t) + C(I_t)]$$

$$\text{s.t.} \quad \sum_{j=1}^{N} X_{jt}\lambda_{jt} \leq x_t; \sum_j R_{jt}\lambda_{jt} \leq R_t \qquad (5.1.23)$$

$$\sum_j Y_{jt}\lambda_{jt} \leq y_t; y_t \leq \overline{y}_t$$

$$I_t = I_{t-1} + y_t - d_t; t = 1, 2, ..., T$$

$$\lambda_t' e = 1, \lambda_t \geq 0, I_0 = I_T = 0; I_t \geq 0$$

Here $C(y_t)$ and $C(I_t)$ are the production and inventory costs and it is assumed that demand $(d_t)$ cannot be backlogged. Clearly we need the condition

$$\sum_{t=1}^{\tau} \overline{y}_t \geq \sum_{t=1}^{\tau} d_t, 1 \leq \tau \leq T$$

for a feasible and hence an optimal solution to exist. This generalized model (5.1.23) is closely related to the optimal production scheduling model known as the HMMS model in operations research literature, where the cost functions $C(y_t)$, $C(\overline{y}_t)$, $C(I_t)$ are usually assumed to be quadratic in form. The only difference is that it is modified so as to include the DEA framework for comparative performance. Taking the above cost functions in a linear form, one could determine from the above LP model a time profile of the operating cost frontier having several components of efficiency, e.g., cost efficiency due to inputs, outputs, capacity and inventory. Sengupta (1998a) has discussed a simplified version of this model and compared it with the HMMS model due to Holt, Modigliani, Muth and Simon (1960).

Two general comments are in order with regard to separating the market efficiency from cost efficiency. First, the empirical studies of technical or production efficiency of competitive firms by DEA models have shown that gross profit margins help explain less than 40 percent of DEA-type technical efficiency. Thus a recent study of 14 international airlines by Sengupta (1998b) shows the following regression results.

$$\ln \theta^* = \underset{(t=28.0)}{4.361} + \underset{(1.99)}{0.063} z_1 + \underset{(0.19)}{0.008} z_2 ; R^2 = 0.320$$

$$\ln \theta^* = \underset{(89.9)}{4.391} + \underset{(2.16)}{0.064} z_1 \qquad\qquad R^2 = 0.320$$

where $\theta^*$ is the DEA efficiency measured as defined in (5.1.1) and $z_1$, $z_2$ are the gross profit margin and volume of passenger demand in log units.

A second point is that fluctuations in demand (or prices) or unit costs due to stochastic factors cannot be easily captured in linear forms; hence quadratic cost functions representing variance terms may have to be explicitly utilized in the DEA model; otherwise biased measures of efficiency may be generated.

## 5.2 Electric Power Industry: Cost and Scale Efficiency

Recent empirical studies of DEA models and their applications have not paid much attention to the statistical distribution of efficiency which may be defined either as extra output (i.e., excess of optimal over actual output), or as input saving, due to the adoption of the best practice production function. Farrell and Fieldhouse (1962) in their empirical estimation of the efficient production function from cross-section data of agricultural farms paid a great deal of attention to the distribution of efficiency across farms. They followed a grouping method, which consists simply in grouping the observations according to output and then estimating the efficient production function for each output group separately by a sequence of linear programming (LP) models. Empirically the frequency distribution of output efficiency was found to be J-shaped in form, which would follow from a beta probability density. Farrell and Fieldhouse (1962) however did not fit any frequency curve of a specific form, nor determined a curve of 'best fit'. Yet such a procedure would be greatly helpful in applied work for two reasons. First of all, the curve of best fit would provide an approximation to the underlying distribution of efficiency, which characterizes the variations in efficiency leading to deviations of observed output from their efficient levels. Thus a probabilistic interpretation of efficiency would be possible, e.g., setting up an approximate confidence interval. Secondly, the method could be more profitably applied to estimate an efficient cost function, rather than a production function. Hence the efficient cost function (e.g., cost frontier) measures the "price" or "allocative efficiency" which measures a firm's success in choosing an optimal set of inputs with a given set of input prices so as to minimize the total factor costs. Farrell (1957) noted the reliability of the cost frontier estimates by the LP method, since cost data are more homogeneous than output and also more accurate in some circumstances.

Our objective here is to analyze the cost efficiency of power plants in the U.S. electric utility industry in terms of the efficiency distribution approach initiated by Farrell (1957) and others. Our empirical study has two basic motivations. One is to develop a nonparametric test procedure for comparing the alternative forms of the efficiency distribution, which result from Farrell's LP approach. Our empirical study attempts to develop and apply a systematic statistical procedure with appropriate parametric and semiparametric tests.

This is done by a two-stage method of semiparametric estimation. In the first stage we apply Farrell's linear programming approach to identify the subset of cost efficient units in the whole sample and derive the empirical probability distribution of costs in a parametric form. The second stage then utilizes the empirical distribution function in order to develop an econometric estimate of the cost frontier. A second objective is to test the existence of economies of scale for the power industry data, which were previously analyzed by Nerlove (1963) and Greene (1990). For this purpose we compare the Farrell method of estimation with the ordinary least squares (OLS).

Some motivations for this two-stage approach may be added here. First of all, Farrell's programming approach is more specific in identifying the efficient units, hence it is used to derive the first subsample $S_1$ of efficient units only. We concentrate on the distribution of the efficient units only in identifying the parametric form. Once this is identified, then we consider the whole sample S comprising both $S_1$ and $S_2$ subsamples in order to estimate the parameters of the cost frontier. Secondly, the cost distribution estimated in the first stage is useful in other respects. For one thing one could use some convenient approximations for the empirical distribution and set up confidence intervals for the mean or the median. Secondly, a mixture of the two distributions belonging to the subsets $S_1$ and $S_2$ may be formulated, where $S_1$ contains the cost-efficient units and $S_2$ the rest. If $\pi_1$ and $\pi_2 = 1—\pi_1$ are the proportions in the two subsets and $F_1(\varepsilon)$, $F_2(\varepsilon)$ are the corresponding distributions, then the mixture distribution $F(\varepsilon) = \pi_1 F_1(\varepsilon) + \pi_2 F_2(\varepsilon)$ provides a convenient model for analyzing the role of the efficient units measured by the proportion $\pi_1$. For example, if $\pi_1$ is very high, say 0.90 or more, the efficient units will have a more dominant role. The reverse case is when $\pi_2 \geq$ 0.90. The mixture distribution model helps to explain an anomaly, which was observed by Farrell and Fieldhouse (1962) in the form of bimodal distribution curves in their empirical studies of production functions.

Finally, our approach is somewhat different from the composite error model followed in econometric estimation. In the composite error model it is usually assumed that the efficient units have a normal distribution, while the inefficient units have one-sided errors. Since our empirical estimates of the efficient units of the subsample $S_1$ do not satisfy the normality assumption, we did not follow the composite error model structure.

## A. Empirical Data

The empirical data over which we apply the efficient distribution approach of Farrell are taken from the cross-section sample of firms in the U.S. electric utility industry available in Greene (1990). 'This data set for 1976 comprises 123 individual firms (i.e., steam plants) with their outputs measured in kwh, three input prices (e.g., of labor, capital and fuel) and total input costs measured in constant dollars. Data on expenditures for labor and fuel used in steam plants for electric power generation are available by firm in the published statistics of the U.S. Federal Power Commission but the capital costs of production are not directly available. We have followed Nerlove's (1963) method for estimating the capital costs of production, i.e., this was done by taking interest and depreciation charges on the firm's entire production plant and multiplying by the ratio of the value of steam plant to total plant as carried on the firm's balance sheets. Wage rate data available from the reports of the Bureau of Labor Statistics were used as price of labor. For fuel, prices were taken on a per-btu basis, since coal, oil or natural gas may be used to produce the steam required for steam electricity generation. These statistics are available in the Statistical Bulletin published by the Electric Power Research Institute.

We had two motivations for choosing this data set. One is that it provides a benchmark for comparison with the parametric approach for estimating the cost frontier attempted by Nerlove (1963) and more recently by Greene (1990). The second motivation is that the electricity generation in U.S. by private companies included in the data set has market characteristics, where the input prices can be considered exogeneous. This is so because output prices are set by the Public Utility Commissions, wage rates by long term contracts after competitive bargaining and the prices of fuel and capital are competitive with other users. We may also note that the industry showed significant scale economies in the previous OLS estimates by Nerlove (1963) and Greene (1990).

The major shortcoming of the data is the way the proxy variable for capital costs of production was constructed to estimate the price index for capital. For many reasons it is well known that depreciation and interest charges do not adequately reflect capital costs in their economic sense. Moreover the depreciation practices vary from firm to firm and such variations introduce an element on non-comparability to some extent. Similarly the estimate of interest cost obtained by taking the current yield on the firm's most recently issued long term bonds has an element of arbitrariness due to the heterogeneity of the capital structure. In spite of these shortcomings the cost frontier provides useful insights about the sensitivity of p0roduction costs in relation to the scale of output and the other input prices such as the wage rate and fuel price.

## B. Modelling Issues

A production frontier implies by the duality theorem a specific factor price or cost frontier. This frontier can then be defined and estimated so as to derive the optimal input set which minimizes total costs. Under competitive market conditions where the input and output price data are available, the estimation of efficiency by a cost rather than a production frontier is much easier for two reasons. For one thing, costs are more homogeneous in that they can be aggregated or disaggregated easily; also they introduce more directly the allocative efficiency problem for the competitive firm, where the allocative or price efficiency measures a firm's success in choosing an optimal set of inputs with a given set of prices. Secondly, the profit function becomes unbounded, when the production function exhibits increasing returns to scale and in such situations a production frontier cannot be specified under the first order conditions of profit maximization. A cost frontier can however be defined in a conditional sense for any given output level. Thus if y is the given level of output and the production function is $f(x) = f(x_1, x_2, \ldots, x_m)$ then the cost minimization model takes the form

$$\text{Min } c(x) = \Sigma \, q_i x_i$$
$$\text{s.t.} \quad y \leq f(x_1, x_2, \ldots, x_m) \tag{5.2.1}$$

where $q_i$ are the input prices for the m inputs. This model then yields the cost frontier:

$$c^* = c^*(q_1, q_2, \ldots, q_m; y) \tag{5.2.2}$$

Recently, production frontiers have been estimated by nonparametric methods, thanks to the seminal paper by Farrell (1957), who developed two efficiency measures at the firm level, e.g., technical efficiency and price (or allocative) efficiency. The first defines the production frontier which measures the firm's success in producing maximum output from a given set of inputs and the second defines the allocative efficiency measuring the firm's success in choosing an optimal set of inputs with a given set of prices. While nonparametric techniques based on Farrell's convex hull method have been frequently applied in estimating production frontiers (see, e.g., Sengupta 1989, 1990), cost frontiers have been usually estimated by parametric methods. Thus Greene (1990) has estimated a stochastic cost frontier model in a loglinear form where the additive error term is assumed to follow a gamma distribution. Based on the cross-section data of electric power plants in the U.S., this parametric model tests for the existence of increasing returns to scale. The assumption of gamma distribution for the error term is however ad hoc and not based on any empirical considerations. The specification of a loglinear production function $f(x)$, also known as the Cobb-Douglas function, has been frequently used in applied economics to test the existence of economies of scale. By the duality theory it leads to a loglinear cost frontier. Thus if the production function $f(x)$ in (5.2.1) is of a Cobb-Douglas form

$$\ln y = \alpha_0 + \sum_{i=1}^{m} \alpha_i \ln x_i$$

then the cost frontier is of a loglinear form

$$\ln c^* = k_0 + \sum_{i=1}^{m} k_i \ln q_i + (1/r)\ln y^* \qquad (5.2.3)$$

where $k_i = \alpha_i/r, r = \sum_{i=1}^{m} \alpha_i$ and $k_0$ is a constant intercept depending on the parameters $\alpha_0, \alpha_1, \ldots, \alpha_m$. Clearly any observed cost ($c_j$) for the j-th firm would then satisfy the specification:

$$\ln c_j = \ln c_j^* + v_j; v_j \geq 0 \qquad (5.2.4)$$

since the cost frontier ($\ln c_j^*$) exhibits minimal costs. If $v_j$ is zero then the firm j is allocatively efficient. If not, then $v_j$ takes a positive value and the firm has a cost level higher than the minimum possible.

In our empirical application the input price of capital ($q_2$) and labor ($q_3$) are expressed relative to fuel price ($q_1$) and then the cost frontier in logarithmic units is expressed as

$$C_j = \beta_0 + \beta_1 Y_j + \beta_2 X_{2j} + \beta_3 X_{3j} + v_j \qquad (5.2.5)$$
$$v_j \geq 0; j = 1, 2, \ldots, N$$

where $C_j$ and $Y_j$ are total costs and output in logarithm units for firm j and $X_2, X_3$ denote the relative price ratios $q_2/q_1$ and $q_3/q_1$ respectively in log units. As Nerlove (1963) has shown that using relative prices as regressors helps specify the presence of economies of scale uniquely. Denoting $Y_j$ by $X_{1j}$ and $C_j$ by $z_j$ we can express the cost frontier equation (5.2.5) in a convenient linear form as

$$z_j = \sum_{i=0}^{3} \beta_i X_{ij} + v_j; v_j \geq 0$$

where $X_{0j} = 1$ for all j. Now for the parametric approach we need to assume a specific distribution for the nonnegative error term $v_j$. Greene (1990) assumed a gamma distribution for the nonnegative error term and applied a nonlinear maximum likelihood method of estimation. Others have used half-normal and exponential distributions for the error term. None of these methods however are data based and hence are ad hoc.

We consider here a nonparametric approach that is entirely data based. This approach initially proposed by Farrell (1957) adopts a two-step method of estimating a frontier. In the first step, each observed unit (i.e., firm k for instance) is tested for efficiency by running a linear programming (LP) model as:

$$\text{Min } v_k = -\beta'X_k \text{ subject to } z \geq X'\beta; \beta \geq 0 \tag{5.2.6}$$

Here prime denotes transpose and $X_k$ is the input vector for the k-th unit (firm). Clearly the unit k is efficient, i.e., it is on the cost frontier if it satisfies for the optimal solution vector $\beta* = \beta*(k)$ on the LP model (5.2.6) the conditions: $z_k = z_k^* = \beta*' X_k$ and $s_k^* = z_k - z_k^* = 0$, where $s_k^*$ is the optimal value of the slack variable. The condition that the optimal slack variable is zero has been introduced in the recent literature on data envelopment analysis (see, e.g., Sengupta, 1989) in order to prevent degeneracy of the optimal solution. Moreover the unit k is inefficient if the above optimality conditions fail to hold, which implies that $z_k > z_k^*$, i.e., observed cost is higher than the minimum level obtainable. At the second step we vary k in the objective function over the index set $I_N = \{1,2,...,N\}$ of all observed units and thus determine two subsets $S_1$, $S_2$ of efficient ($S_1$) and inefficient ($S_2$) units. Once the units in the efficient subset $S_1$ are determined one can apply two methods for determining the final estimate of the cost frontier: one parametric and the other nonparametric. For the parametric case we derive the empirical form of the cost distribution for all units belonging to $S_1$ and then estimate the parameters of the distribution, e.g., by the generalized method of moments, which has been applied recently by Kopp and Mullahy (1990). In the second case we may apply either a semiparametric or a nonparametric method. One semiparametric method often used in applied studies is based on the median rather than the mean. This adopts the criterion of the least sum of absolute deviation (LAD) of errors over all units $k \in S_1$, which yields the LP model:

$$\text{Min}_{\beta} \sum_{k \in S_1} |v_k| \text{ subject to } z_j \geq \beta'X_j; \beta \geq 0$$
$$j,k \in S_1.$$

Since $v_j = z_j - \beta'X_j$ for all $j \in S_1$ is nonnegative, this reduces to the following:

$$\text{Max}_{\beta} \overline{g} = \beta'\overline{x} = \sum_{i=0}^{3} \beta_i \overline{x}_i$$
$$\text{subject to} \qquad \beta \in C(\beta) \tag{5.2.7}$$

where

$$C(\beta) = \{\beta | z \geq \beta'X; \beta \geq 0; j \in S_1\}$$

$$\bar{x} = (\bar{x}_i), \bar{x}_j = (1/N_1)\sum_{j=1}^{N_1} X_{1j}; j \in S_1$$

This type of LAD model has been frequently used in applied economics and DEA literature, e.g., Timmer (1971) and Sengupta (1989). Koenker (1987) has compared the relative efficiency of the LAD estimates with ordinary least squares.

In the nonparametric case we estimate the probability density $p_H(z)$ of $z_j$ over the subset $S_1$ which includes the units found to be efficient by Farrell's method. We partition the closed interval $[a,b]$ of z by $a = z_{(0)} < z_{(1)} < z_{(2)} < z_{(k)} = b$ and consider the histogram estimates of the form

$$p_H(z) = \begin{cases} c_r & \text{for } z_{(r)} < z < z_{(r+1)}, r = 0,1,...,k-1 \\ c_{r-1} & \text{for } z_{(k)} = b \\ 0 & \text{otherwise} \end{cases} \tag{5.2.8}$$

where $p_H(z) > 0$ and $\int_a^b p_H(z)dz = 1$. To estimate the population histogram of this form we consider the entire sample space in subset $S_1$ and count the number of observations falling in the r-th interval. Let $N_r$ be this number. Then the population parameter $c_r$ above can be estimated by $\hat{c}_r = N_r[N(z_{(r+1)} - z_{(r)}]^{-1}$ for r=0,1,...,k-1. The same estimates can be obtained also by maximum likelihood, if one assumes that the samples are generated by a multinomial distribution. As Eubank (1988) has shown that these estimates are statistically consistent under large sample conditions, even when the true population density is unknown.

Our generalization of Farrell's efficiency distribution method involves three aspects. The first consists in determining the subset $S_1$ of efficient units by applying Farrell's LP model to each of the N units. We then estimate the empirical distribution of optimal costs over the efficient subset $S_1$ by following the method of moments in a generalized form. The generalization involves a set of moment restrictions discussed by Kopp and Mullahy (1990) which provides the basis for identification of the mean technical efficiency parameter $\mu$ defined by the expectation term $E(v_j)$ in Farrell's model (5.2.5). The second feature of the efficiency distribution method involves several approximations of the observed distribution of optimal costs over the subset $S_1$ in terms of the exponential and the normal, which are the two most common distributions frequently applied in models of data envelopment analysis. An important question here is: which of the several approximate distributions fits the observed efficiency data best in the sense of a goodness of fit criterion. We apply both chi-square and Kolmogorov-

Smirnov tests to determine the best fitting distribution. Finally, we compare the estimates of the cost frontier by the LAD method applied over the efficient subset $S_1$ defined by (5.2.7) and the OLS estimates. The reason for including the OLS method is the fact that it is closely related to the corrected OLS method, which transforms the cost frontier equation (5.2.5) as:

$$z_j = (\mu + \beta_0) + \sum_{i=1}^{3} \beta_i X_{ij} + \varepsilon_j; \varepsilon_j = v_j - \mu$$
$$\text{where } \mu = E(v_j), v_j \geq 0, j = 1, 2, ..., N$$

Since the estimate of the scale coefficient is invariant under both OLS and COLS, it is convenient to compare the LAD and OLS estimate of scale.

## C. Empirical Results

Based on the efficient units determined by Farrell's LP model (5.2.6) we first identify the probability density function $f(z)$, which characterizes the optimal cost data most closely. Here we restrict ourselves to the Pearson system of curves and its generalization by Elderton and Johnson (1969). The generalized method of moments is applied to the subset $S_1$ of 123 firms and this yields a J-shaped beta density function, which is a Type I curve in Pearsonian system as follows:

$$f(z) = 138.80(1 + 6.289z)^{-0.074}(1 - 0.884z)^{5.564}$$

where $z_j$ denotes optimal cost in logarithmic units. To see if these moment estimates can be improved further in terms of estimation, we also considered the maximum likelihood method but since this involved the solving of highly nonlinear equations involving digamma and trigamma functions we did not proceed further. Instead we followed a simpler method of approximations, by which we assumed the optimal cost data $(z_j, j \in S_1)$ to follow two other possible densities: the exponential and the normal. These empirical density functions appear as follows:

Exponential $f(z) = 5.882 \exp(-5.882z)$
Normal $f(z) = (47.62 / \sqrt{2\pi}) \exp[0.5(47.62z - 8.10)^2]$

One may also note that the LAD model (5.2.7) yields another distribution of optimal costs, which is also estimated by the nonparametric method outlined in (5.2.8) before. These empirical distribution functions are reported in Table 5.2.3.

The information contained in these distributions can be utilized in several ways. First of all, one could test if the normal approximation (or for that matter half normal distribution) is any better than the exponential. For this purpose one could compare the two approximations by chi-square and Kolmogorov-Smirnov

(KS) statistics, where the first is a parametric test and the second nonparametric. The numerical results are as follows:

|              | Exponential vs. beta | normal vs. beta |
|--------------|:---:|:---:|
| Chi square   | 0.0320 | 5.1252 |
| KS statistic | 0.0355 | 0.5900 |

Since smaller values indicate better approximation, it is clear that the exponential density provides a better approximation. Since the exponential is a special case of gamma density where the ML method can be directly applied, a parametric cost frontier model based on gamma distribution would be most appropriate here. Greene (1990) used this as an assumption in his cost frontier model, but in our case it is derived from the empirical estimates of Farrell's cost frontier. Secondly, one could compare the beta distributions of costs for the two sample sets $S_1$ and $S_2$ defined before and test if they are statistically different. The numerical result here reported in Table 5.2.1 yields a chi square value exceeding 6.21 and the KS statistic of 0.71 suggesting a significant difference between the two output distributions. In economic terms this implies that if the data on individual units are first stratified into efficient and non-efficient units and then estimated, one would obtain a more reliable estimate of the cost frontier. Thirdly, one could directly compare the LAD estimates of the cost frontier with the OLS estimates, which are based on approximate normality. This comparison is reported in Tables 5.2.2 and 5.2.3. Finally, the distributions underlying the LAD and OLS estimates above can be directly compared in terms of the stochastic dominance criteria, which are essentially nonparametric in character. Let x and y be two output distributions. Stochastic dominance of one profile y over another x indicates that y is preferred to x by the expected utility maximizing agents. Thus strict stochastic dominance of first (FSD) and second order (SSD) are defined as follows:

(FSD):  p FSD q if $\int u(v)\,dp(v) > \int u(v)\,dq(v)$
 for every increasing utility function $u(v)$ and

(SSD):  p SSD q if $\int u(v)\,dp(v) > \int u(v)\,dq(v)$
 for every increasing and strictly concave utility function $u(\cdot)$

where p and q are the distributions for y and x respectively.

Clearly in case of costs the above inequalities are reversed, i.e., if p FSD q then the expected loss under q is less than that under p implying that q will be preferred by an expected cost minimizing agent.

For statistical testing of the stochastic dominance of first and second order, one could apply the Kolmogorov-Smirnov one-sided two-sample statistic that can be used to test the null hypothesis $p(v) = q(v)$ against the alternative $q(v)$

$> p(v)$. If $F_N$ and $G_N$ are two empirical distribution functions based on N samples, then the statistic

$$D^+ = \sup_{-\infty < v < \infty} [G_N(v) - F_N(v)]$$

is the one-sided Kolmogorov-Smirnov (KS) statistic whose distribution has been numerically tabulated in Lilliefors (1967). Results of these tests are summarized in Table 5.2.4, which show that LAD estimates are significantly different from the OLS (normal) estimates and one cannot statistically distinguish between the exponential and the beta distribution.

We may now briefly comment on the implications of various statistical results. First of all, the J-shaped beta density appears to represent in our empirical case the cost distribution of the Farrell model most adequately. We note that for their farm management data Farrell and Fieldhouse (1962) found in their grouping method this type of J-shaped frequency distribution of efficiencies, although they plotted it only in figures. This beta distribution can be used to characterize the confidence interval for the mean technical efficiency in Farrell's model. Also nonlinear ML methods can be applied to improve upon the estimates by the method of moments. Secondly, the comparison of the LAD and OLS estimates in Table 5.2.2 show that the former are statistically significant and also stable when tested by their t-ratios. Since the LAD estimates are closely related to median unbiasedness, they have a feature of robustness, which has been confirmed by other simulation studies, e.g., Dielman and Pfaffenberger (1982). Note that we have the evidence of significant scale economies in both LAD and OLS models. The return to scale however is much lower in the LAD model. Since the underlying distribution of optimal costs follows a nonnormal distribution, the LAD estimates are more likely to outperform the OLS and corrected OLS estimates in terms of statistical efficiency. Finally, the stochastic dominance test results reported in Tables 5.2.3 and 5.2.4 show that the LAD distribution is the most dominated in the first order sense by other competing distributions. This implies that the rational cost minimizing agents would prefer this estimate over others. Thus, by using the exponential distribution for the error terms in the cost frontier model (5.2.5) one could apply the nonlinear ML method of estimation.

| Efficiency (in log units of cost) class midpoints | Observed Frequency (%) | Exponential (%) | Beta (%) | Normal (%) |
|---|---|---|---|---|
| >0.072 | 0.068 | 0.347 | 0.359 | 0.0 |
| 0.108 | 0.186 | 0.228 | 0.241 | 0.014 |
| 0.179 | 0.186 | 0.150 | 0.158 | 0.985 |
| 0.250 | 0.153 | 0.099 | 0.101 | 0.001 |
| 0.321 | 0.153 | 0.065 | 0.0.63 | - |
| 0.329 | 0.085 | 0.044 | 0.037 | - |
| 0.463 | 0.068 | 0.019 | 0.022 | - |
| 0.534 | 0.051 | 0.019 | 0.011 | - |
| 0.605 | 0.034 | 0.012 | 0.005 | - |
| >0.640 | 0.016 | 0.008 | 0.003 | - |
| Sum | 1.00 | 1.00 | 1.00 | |

Table 5.2.1. Distribution of efficiency and various fitted curves

| | Farrell (LAD) | t-ratio | OLS | t-ratio |
|---|---|---|---|---|
| $\beta_0$ (intercept) | -2.623 | -1.15 | -3.49 | -3.26** |
| $\beta_1$ | 0.259 | 25.90** | 0.225 | 9.03** |
| $\beta_2$ | 0.118 | 1.87* | 0.411 | 1.98* |
| $\beta_3$ | 0.586 | 9.45** | 0.335 | 1.63 |
| $R^2$ | | | 0.623 | |
| App. $R^2$ | 0.265 | | - | |
| Cost elasticity | 0.259 | | 0.225 | |
| Return to scale | 3.861 | | 4.444 | |

Table 5.2.2. Cost frontier parameter estimates

Notes:
1. LAD t-ratio are based on bootstrap standard error computed by the algorithm of Hardle and Bowman (1988).
2.   App. $R^2$ for the Farrell estimates denotes an approximate measure of $R^2$ defined as $[1-(\Sigma s_i^2 / \Sigma[c_i - med(c)]^2$ where s is the slack variable, c = cost and med(c) is median cost.
3.   One and two asterisks denote significant t-values at 5% and 1% respectively.

A. Estimates of cumulative distributions

| Class Midpoints | Beta cdf I | Exponential cdf II | Lad cdf III | Normal cdf IV |
|---|---|---|---|---|
| 0.072 | 0.3594487 | 0.347418924 | 0.2881356 | 3.04425E-10 |
| 0.108 | 0.5999537 | 0.575124902 | 0.4576271 | 0.014215606 |
| .179 | 0.7583206 | 0.725246686 | 0.7288136 | 0.999044664 |
| .250 | 0.8597509 | 0.024218835 | 0.8474576 | 0.999999998 |
| .321 | 0.9225419 | 0.8894691 | 0.8813559 | 0.999999998 |
| .392 | 0.9598399 | 0.932487234 | 0.9152542 | 0.999999998 |
| .463 | 0.9809059 | 0.960848188 | 0.9661017 | 0.999999998 |
| .534 | 0.9920854 | 0.97954972 | 0.9830508 | 0.999999998 |
| .605 | 0.9975698 | 0.991873029 | 0.9830508 | 0.999999998 |
| >0.640 | 1 | 1 | 1 | 0.999999998 |

B. Values of Kolmogorov-Smirnov (KS) statistics

| | I-II | I-IV | III-I | III-II | III-IV |
|---|---|---|---|---|---|
| 0.072 | 0.0120298 | 0.35944872 | -0.0713131 | -0.05928333 | 0.288135593 |
| .108 | 0.0248288 | **0.585738094** | **-0.1423266** | **-0.11749778** | **0.443411512** |
| .179 | 0.0330739 | -0.240724082 | -0.029507 | 0.003566873 | -0.270231104 |
| .250 | 0.035532 | -0.140249125 | -0.0122932 | 0.023238793 | -0.152542371 |
| .321 | 0.0330728 | -0.07745811 | -0.041186 | -0.00811317 | -0.118644066 |
| .392 | 0.0273526 | -0.040160116 | -0.0445856 | -0.017233 | -0.084745761 |
| .463 | 0.0200577 | -0.019094081 | -0.0148042 | 0.005253507 | -0.033898303 |
| .534 | 0.0125394 | -0.007914644 | -0.0090345 | 0.003504875 | -0.16949151 |
| .605 | 0.0056968 | -0.00243017 | -0.0145919 | -0.00882218 | -0.016949151 |
| >0.640 | 0 | 1.83068E-09 | -2.22E-16 | -2.2204E-16 | 1.83068E-09 |

Table 5.2.3. Comparative Kolmogorov-Smirnov Statistics for Distributions of Inefficiency

Note: Bold numbers are the KS test statistic

| Midpoint (scaled) | Beta vs. Exponential | LAD vs. Normal |
|:---:|:---:|:---:|
| 0.036 | 0.012 | 0.288 |
| 0.108 | 0.025 | 0.443 |
| 0.179 | 0.033 | -0.270 |
| 0.250 | 0.035 | -0.152 |
| 0.321 | 0.033 | -0.119 |
| 0.392 | 0.027 | -0.085 |
| 0.463 | 0.020 | -0.034 |
| 0.534 | 0.012 | -0.017 |
| 0.605 | 0.006 | -0.017 |
| 0.676 | 0.0 | 0.0 |
| | | |
| KS statistics | 0.035 | 0.443** |

Table 5.2.4. Kolmogorov-Smirnov Statistics for Comparing Cost Distributions[a]

[a]The sample set here includes only those units found to be efficient by Farrell's method.
**Denotes significance at 1% level.

## 5.3 Airlines Industry: Competition and Efficiency

World airlines industry captures in many ways the interplay of the various economic forces of market competition to a significant degree. First of all, air passenger traffic according to the estimate by Oum and Yu (1998) increased throughout the 1980s at an average rate of 6% per year, while world GDP growth averaged about 3% during the same period. This implies an income elasticity of 2.0 for air travel demand. However, there is a strong pro-cyclical relationship between world economic growth and the total air traffic demand including both passenger traffic and freight. According to IATA's forecast, airlines based in Asia-Pacific economies are likely to exhibit faster growth than airlines in other countries. Thus, given a forecasted average of 7.2% annual growth, Asian carriers on average will double in size in 10 years, while the North American carriers can expect an average growth of 50% in the next 10 years. Secondly, cost competitiveness determined by production efficiencies and input costs (prices), can be directly measured by the allocative DEA model, which minimizes total overall costs subject to the production constraints. Oum and Yu (1998) have analyzed of time series data of 22 world airlines for the period 1980-1995 divided into three groups: North American carriers, European carriers and Asian carriers. They estimated stochastic production and cost frontiers by nonlinear maximum likelihood, which provides average estimates of production (technical) and cost efficiency. The same data are used here to estimate nonparametric efficiencies by data envelopment analysis. This DEA application provides firm-specific efficiencies. Finally, the cross-section sample of international airlines provides a direct test of the hypothesis that the competitive pressure of market demand and global trade tend to improve the scale economies and cost efficiency of individual airlines competing in the world market.

We consider here two empirical applications. One is the DEA application of a cost competitiveness model involving input output data set of 14 airlines averaged over the period 1988-90 taken from Schefczyk (1993). The second DEA application uses the time series data set (1986-95) from Oum and Yu (1998) and computes the technical and allocative efficiency of 22 world airlines divided into three groups. We also compare the efficiency persistence over time of specific airlines for the period 1980-95.

In the first application three input costs ($x_i$) and two output revenues ($y_r$) are considered as follows: $x_2$ = operating cost defined as total operating expenses minus rent, depreciation and amortization, $x_3$ = cost of total nonflight assets, $y_1$ = passenger kilometer revenue and $y_2$ = nonpassenger ton-kilometer revenue. Besides these input costs and revenues, the other instrument variables which directly affect airlines efficiency are the following, that are reported by Schefczyk: $z_1$ = gross profit margin and $z_2$ = international passenger load.

Table 5.3.1 presents the nonparametric estimates of cost efficiency $\theta_k^*$ for the 14 airlines with 6 efficient and 8 inefficient. However, since one airline (e.g., Iberia) attains the level 0.999 which can be rounded to 1.00, this one may be included in the efficient set $S_1$, in which case half of the total is efficient, the other half being inefficient. The least efficient airline is AU Nippon with a value of efficiency score $\theta^* = 0.844$. This implies that this airline would have to reduce its input costs to 84.4 percent of the current level to become efficient.

To compare different airlines belonging to the efficient set $S_1$, determined by model (5.1.2) where each has efficiency score $\theta^* = 1.0$, we estimate the optimal value $\alpha_0^*$ given by the dual LP model (5.1.7) and the results are as follows:

| Airline | $\alpha_0^*$ | Returns to scale |
|---|---|---|
| Cathay Pacific | 0.426 | IRS |
| Lufthansa | 0.015 | IRS |
| Singapore | 1.264 | IRS |
| Korean Air | 0.0 | CRS |
| Quantas | 0.327 | IRS |
| UAL | 0.0 | CRS |
| Iberia | 1.0 | IRS |

Note that a positive value of $\alpha_0^*$ indicates the presence of increasing returns to scale (IRS), zero and negative being constant and diminishing returns. Clearly, Singapore Airlines tops the list in terms of the size of IRS and Lufthansa the least, with Korean Air and UAL displaying CRS.

| Airline | θ* Efficiency Score | Input $x_1$ Actual | Optimal | Input $x_2$ Actual | Optimal | Input $x_3$ Actual | Optimal |
|---|---|---|---|---|---|---|---|
| Air Canada | 0.893 | 5,723 | 5,111 | 3,239 | 2,892 | 2,003 | 1,788 |
| AU Nippon | 0.844 | 5,895 | 4,975 | 4,225 | 3,566 | 4,557 | 3,846 |
| American | 0.948 | 24,099 | 22,846 | 9,560 | 9,063 | 6,267 | 5,941 |
| British Air | 0.959 | 13,565 | 13,008 | 7,499 | 7,191 | 3,213 | 3,081 |
| Cathay Pacific | 1.000 | 5,183 | 5,183 | 1,880 | 1,880 | 783 | 783 |
| Delta | 0.977 | 19,080 | 18,641 | 8,032 | 7,847 | 3,272 | 3,197 |
| Iberia | 0.999 | 4,603 | 4,598 | 3,457 | 3,453 | 2,360 | 2,358 |
| Japan | 0.859 | 12,097 | 10,391 | 6,779 | 5,823 | 6,474 | 5,561 |
| KLM | 0.973 | 6,587 | 6,409 | 3,341 | 3,251 | 3,581 | 3,484 |
| Korea Air | 1.000 | 4,654 | 5,654 | 1,878 | 1,878 | 1,916 | 1,916 |
| Lufthansa | 1.000 | 12,559 | 12,559 | 8,098 | 8,098 | 3,310 | 3,310 |
| Quantas | 1.000 | 5,728 | 5,728 | 2,481 | 2,481 | 2,254 | 2,254 |
| Singapore | 1.000 | 4,715 | 4,715 | 1,792 | 1,792 | 2,485 | 2,485 |
| UAL Corporation | 1.000 | 22,793 | 22,793 | 9,874 | 9,874 | 4,145 | 4,145 |

Table 5.3.1 Cost efficiency based on the DEA model

Note: $x_1$ = available ton kilometer, $x_2$ = operating cost, $x_3$ = nonflight assets

| Sample | Intercept | $x_1$ | $x_2$ | $x_3$ | $x_4$ | $\bar{R}^2$ |
|---|---|---|---|---|---|---|
| Total | -49150 | 6.91 | -3.62 | 0.39 | 63.44 | 0.961 |
| (N = 14) | (t=0.86) | (7.86) | (-1.71) | (0.22) | (0.82) | |
| Efficient set | -78519 | 8.32 | -7.12 | 5.48 | 87.80 | 0.902 |
| ($N_1$ = 6) | (-0.39) | (2.94) | (-0.99) | (0.57) | (0.33) | |
| Inefficient set | 61609 | 4.24 | 1.17 | 0.55 | -92,53 | 0.977 |
| ($N_2$ = 8) | (0.62) | (1.81) | (0.24) | (0.25) | (-0.63) | |
| Total | -4301 | 5.20 | - | - | - | 0.956 |
| (N = 14) | (-1.12) | (16.92) | | | | |
| Efficient set | -4987 | 5.23 | - | - | - | 0.902 |
| ($N_1$ = 6) | (-0.57) | (6.87) | | | | |
| Inefficient set | -3601 | 5.16 | - | - | - | 0.986 |
| ($N_2$ = 8) | (-1.18) | (22.39) | | | | |
| Total | -64292 | 5.59 | - | - | 80.63 | 0.957 |
| (N = 14) | (-1.13) | (11.76) | | | (1.06) | |
| Total | -70873 | 5.76 | - | -0.86 | 91.56 | 0.954 |
| (N = 14) | (-1.17) | (9.36) | | (-0.47) | (1.11) | |

Table 5.3.2 Regression estimates of the linear production function (dependent variable: $y_1$)

Note:   $\bar{R}^2$ denotes adjusted $R^2$, adjusted for degrees of freedom.

| Dependent Variable | Inputs | | | | | Instrument variables | | R² |
|---|---|---|---|---|---|---|---|---|
| | Intercept | $x_1$ | $x_2$ | $x_3$ | $x_4$ | $z_1$ | $z_2$ | |
| θ* | 0.737 (t=1.62) | 1.00E-05 (1.42) | -1.26E-05 (-0.74) | -1.26E-05 (-0.84) | 0.0003 (0.52) | - | - | 0.232 |
| θ* | 0.944 (33.46) | 1.64E-06 (0.73) | - | - | - | - | - | 0.042 |
| log (1000θ*) | 4.361 (27.99) | - | - | - | - | 0.063 (1.99) | 0.008 (0.19) | 0.320 |
| log (1000θ*) | 4.391 (89.92) | - | - | - | - | 0.064 (2.16) | - | 0.317 |
| log $y_1$ | -1.08 (-0.32) | 1.611 (2.013) | -1.371 (-1.551) | 0.894 (1.679) | - | - | - | 0.496 |

Table 5.3.3 Efficiency regression on inputs and other instrument variables.

Note: $z_1$ = gross profit margin, $z_2$ = volume of international passenger demand
$y_1$ = passenger revenue; E-05 = $10^5$

Table 5.3.2 presents a comparative view of production function estimates of the two sets $S_1$, $S_2$ the efficient and inefficient respectively. Here only the most important output $y_1$ is considered as the dependent variable; also $x_4$ is added as an extra explanatory variable representing passenger load factors. Three interesting points come out very clearly. One is that the capacity variable $x_1$ emerges as the major explanatory variable; other explanatory variables $x_2$, $x_3$, $x_4$ have either insignificant coefficients or wrong signs. Secondly, the intercept term for the efficient set $S_1$ is always negative, thus implying IRS in a consistent fashion.

Finally, we have in Table 5.3.3 the estimated results on the possible sources of efficiency, where the efficiency measure ($\theta^*$) or its log equivalent is regressed on the four inputs and two instrument variables $z_1$ and $z_2$ representing gross profit margin and international passenger demand. Only gross profit margin ($z_1$) and capacity input ($x_1$) turned out to be positively correlated with efficiency score, but only the gross profit margin has a significant coefficient at 5% level of t test, when $z_1$ alone is used as the explanatory variable. This suggests that the profit margin alone does not indicate a measure of higher efficiency, i.e., there is a tradeoff of short run profits to other goals like retaining market share and the competitive edge in international air travel market.

Next, we consider the time series data set from Oum and Yu, where five inputs e.g., labor, fuel, materials, flight equipment and ground properties and equipment (GPE) and four outputs e.g., passengers, freight, mail and non-scheduled outputs. For the latest year, 1995 for which the data are available, three types of efficiency: technical, allocative and overall are estimated by the DEA model. Those airlines which, turned out to be efficient in the year 1995, are then tested if they were also efficient in the earlier years. Those which remained efficient for seven out of ten years, are termed persistently efficient.

By the technical efficiency (TE) test only two airlines, e.g., Delta ($\theta^* = 0.9937$) and US Air ($\theta^* = 0.9090$) turned out to be inefficient. When we compute the overall efficiency model specified by the LP model (5.1.4) specified before, the overall efficiency (OE) turned out to be as follows:

|  | North American |  | European |  | Asian |  |
|---|---|---|---|---|---|---|
| Efficient | United | OE=1.0 | Lufthansa: | 1.0 | Japan Airlines | 1.0 |
|  | Northwest | OE=1.0 | British Air: | 1.0 | Singapore: | 1.0 |
|  | Canadian | OE=1.0 | Air France: | 1.0 | Korean | 1.0 |
|  |  |  | KLM: | 1.0 |  |  |
|  |  |  | Swiss Air: | 1.0 |  |  |
| Inefficient | American | OE=0.819 | SAS: | 0.921 | All Nippon: | 0.898 |
|  | Delta | OE=0.976 | Iberia | 0.740 | Cathay Pacific | 0.915 |
|  | US Air | OE=0.663 |  |  | Quantas | 0.662 |
|  | Continental | OE=0.854 |  |  | Thai | 0.627 |
|  | Air Canada | OE=0.885 |  |  |  |  |

Clearly the allocative efficiency (AE) measure can be computed by the ratio $OE_j/TE_j = AE_j$ as follows:

| United | 1.0 | Lufthansa | 1.0 | Japan | 1.0 |
|---|---|---|---|---|---|
| Northwest | 1.0 | British Air | 1.0 | Singapore | 1.0 |
| Canadian | 1.0 | Air France | 1.0 | Korean | 1.0 |
| American | 0.819 | KLM | 1.0 | All Nippon | 0.898 |
| Delta | 0.982 | Swiss Air | 1.0 | Cathay | 0.915 |
| US Air | 0.729 | SAS | 0.921 | Quantas | 0.662 |
| Continental | 0.854 | Iberia | 0.740 | Thai | 0.627 |
| Air Canada | 0.885 |  |  |  |  |

Those airlines for which both TE and AE equal one are in market equilibrium, since they attain the point of tangency of the iso-quant and the iso-cost line. By this test eleven out of 22 airlines do not attain their market equilibrium, i.e., for these inefficient airlines costs can be further reduced thereby improving overall cost efficiency. Hence technical efficiency by itself is not sufficient to generate allocative or overall efficiency. In a dynamic context this disequilibrium framework leads to the so-called learning and adjustment process by the inefficient units.

One way to evaluate the cost of the disequilibrium is to compute the extent of deviations of the individual inputs of each airline from the efficient level computed from the overall efficiency model (5.1.4). The following table shows the surplus or deficit input levels for the five inputs.

|                | Labor | Fuel | Materials | Flight  | GPE     |
|----------------|-------|------|-----------|---------|---------|
| American       | 16448 | 172  | 0.1245    | -0.0006 | 0.3672  |
| United         | 0     | 0    | 0.0       | 0.0     | 0.0     |
| Delta          | 1002  | 142  | 0.0832    | 0.0078  | 0.0989  |
| Northwest      | 0     | 0    | 0.0       | 0.0     | 0.0     |
| US Air         | 14714 | 1.0  | 0.0978    | 0.3104  | 0.0961  |
| Continental    | 4710  | 82   | 0.0434    | 0.0823  | -0.0862 |
| Canadian       | 0     | 0    | 0.0       | 0.0     | 0.0     |
| Japan Airlines | 0     | 0    | 0.0       | 0.0     | 0.0     |
| All Nippon     | 1492  | -199 | 0.2300    | 0.1800  | 0.5620  |
| Singapore      | 0     | 0    | 0.0       | 0.0     | 0.0     |
| Korean         | 0     | 0    | 0.0       | 0.0     | 0.0     |
| Cathay Pacific | 1197  | -166 | -0.1720   | -0.0980 | 0.0340  |
| Quantas        | 7408  | -289 | -0.1080   | -0.0820 | -0.1840 |
| Thai           | 7719  | -505 | 0.1680    | -0.1360 | 0.4020  |
| Lufthansa      | 0     | 0    | 0.0       | 0.0     | 0.0     |
| British Air    | 0     | 0    | 0.0       | 0.0     | 0.0     |
| Air France     | 0     | 0    | 0.0       | 0.0     | 0.0     |
| SAS            | 1417  | 56   | 0.1260    | 0.0820  | -0.1100 |
| KLM            | 0     | 0    | 0.0       | 0.0     | 0.0     |
| Swiss Air      | 0     | 0    | 0.0       | 0.0     | 0.0     |
| Iberia         | 6429  | -82  | -.0890    | 0.0680  | 0.0060  |

Table 5.3.4. Surplus or Deficit Inputs

Note that those airlines which have 100% overall efficiency have no surplus or deficit in input usage.

Another way to look at the OE model is to compare the actual observed cost of all inputs with the optimal costs as follows:

|                | Actual | Optimal |             | Actual | Optimal |
|----------------|--------|---------|-------------|--------|---------|
| American       | 5.74E9 | 4.71E9  | Korean      |        | 5.71E8  |
| United         | 5.29E9 | 5.29E9  | Cathay      | 8.71E8 | 7.97E8  |
| Delta          | 4.23E9 | 4.13E9  | Quantas     | 1.12E9 | 7.41E8  |
| Northwest      | 3.03E9 | 3.03E9  | Thai        | 4.74E8 | 2.97E8  |
| US Air         | 2.87E9 | 1.90E9  | Lufthansa   | 2.67E9 | 2.67E9  |
| Continental    | 1.32E9 | 1.13E9  | British Air | 2.02E9 | 2.02E9  |
| Air Canada     | 0.99E8 | 6.19E8  | Air France  | 2.53E9 | 2.53E9  |
| Canadian       | 5.33E8 | 5.33E8  | SAS         | 1.13E9 | 1.04E9  |
| Japan Airlines | 2.75E9 | 2.76E9  | KLM         | 1.56E9 | 1.56E9  |
| All Nippon     | 1.72E9 | 1.54E9  | Swiss Air   | 1.17E9 | 1.17E9  |
| Singapore      | 4.31E8 | 4.31E8  | Iberia      | 1.20E9 | 8.89E8  |

Table 5.3.5. Actual and Optimal Costs

This shows the extent of cost savings for those airlines which actual costs exceeded the optimal costs e.g., American, US Air, Cathay Pacific, Quantas, Thai and SAS. Clearly by adjustment of the input levels the airlines that are producing under overall inefficiency may improve their performance through optimizing the inputs used.

How does optimal cost vary in response to output? The data on optimal costs obtained from the DEA model can be used in a regression equation to answer this question. Since there is significant multicollinearity among the four outputs, we consider the most important output variable, the passenger demand denoted by $y_1$ and the optimal costs as C. Taking natural logarithms the ordinary least squares (OLS) estimates are as follows:

$$\ln C = 12.895 + \underset{(t=9.19)}{0.964} \underset{(5.818)}{\ln y_1} \qquad R^2 = 0.629$$

$$\ln C = 12.970 + \underset{(8.917)}{0.951} \underset{(5.451)}{\ln y_1} + \underset{(0.321)}{0.008 \, OE^*} \ln y_1 \qquad R^2 = 0.631$$

Both estimates clearly show that passenger demand has a significant impact on cost efficiency. In the second equation the variable OE* denotes a dummy with a value one if the airlines is overall efficient and zero otherwise. Both equations reveal significant scale economies and the cost elasticity of passenger demand exceeds the value 0.950 at a statistically significant level of 1%. The insignificant value of the coefficient for the dummy variable shows that the scale economies do not differ between the efficient and inefficient airlines.

For tests of persistence in terms of overall efficiency over the ten year period 1980-95, only five airlines retained their efficiency persistence e.g., Northwest, Canadian, Continental, Japan Airlines and Singapore Airlines. All other airlines had procyclical variations in overall efficiency.

The regression estimates of Oum and Yu (1998) reached three broad conclusions from their stochastic production functions. First, the Asian carriers enjoyed higher efficiency of labor inputs than North American and European counterparts. This is also confirmed by our DEA results in Table 5.3.4. However, our results are much more robust since they use the nonparametric overall efficiency model. Secondly, the major US carriers were found to be generally more efficient than the Asian carriers, which in turn were more efficient than the European carriers. But this is not confirmed by our DEA results. Whereas our results show technical efficiency at 100% level for 20 our of 22 airlines, the overall efficiency at 100% is attained by only 11 out of 22 airlines which include 3 out of 8 U.S. carriers, 5 out of 7 European carriers and 3 out of 7 Asian carriers. But our results are based on the year 1995. When we consider earlier years our results confirm the finding reached by Oum and Yu that the productivity (efficiency) gap between North American carriers diminished significantly from 1980 onwards. Finally, Oum and Yu's study found the trend in productive efficiency gain to be highly procyclical over the whole period 1986-95. On an overall basis the North American carriers did not achieve any significant productive efficiency improvements during the 1986-93 period. Among North American carriers two airlines e.g., Northwest and Canadian improved productive efficiency the most. Their results on the estimates of average yield and unit cost in current U.S. dollars are as follows:

|      | U.S. Carriers | | European Carriers | | Asian Carriers | |
|------|-------|-----------|-------|-----------|-------|-----------|
|      | Yield | Unit cost | Yield | Unit cost | Yield | Unit cost |
| 1987 | 1.00  | 0.91      | 1.12  | 0.96      | 1.12  | 1.01      |
| 1989 | 1.15  | 1.20      | 1.16  | 1.10      | 1.18  | 1.12      |
| 1991 | 1.18  | 1.21      | 1.37  | 1.33      | 1.21  | 1.17      |
| 1993 | 1.20  | 1.19      | 1.20  | 1.19      | 1.18  | 1.15      |
| 1995 | 1.14  | 1.10      | 1.29  | 1.20      | 1.29  | 1.18      |

These trends are also supported by our DEA results based on the overall efficiency model.

In order to obtain both input-specific and output-specific efficiencies we considered an alternative DEA model for the technical efficiency case as follows

Maximize $\phi - \theta$

s.t.      $$\sum_{j=1}^{N} x_{ij}\lambda_j \leq \theta x_{ik}; \sum_{j=1}^{N} y_{rj}\lambda_j \geq \phi y_{rk} \qquad (5.3.1)$$

$\Sigma\lambda_j = 1; \phi, \theta$ free in sign

This was applied to the data set for 1995. Of the 20 efficient airlines with a TE value of $\theta^* = 1.0$, five turned out to be not efficient in the output-specific sense so that $\phi^* - \theta^*$ is positive. These airlines are Cathay Pacific, Iberia, Japan Airlines and Swiss Air. This implies that the technical efficiency measure is only a partial indicator of productive efficiency. The so-called profiling approach, which tests among the technical efficient units to see if they utilize some of the critical inputs most efficiently or not, yields similar conclusions.

In order to test the adequacy of the technical efficiency measure we perform a loglinear regression test, which regresses the log of the TE measure $\theta^*$ obtained from the DEA model on the five inputs $x_1$ through $x_5$ comprising labor, fuel, materials, flights and GPE

$$\ell n\theta^* = \beta_0 + \sum_{i=1}^{5} \beta_i \ell nx_i + \text{error} \qquad (5.3.2)$$

Another regression used dummy variables to give the results as though one only wanted to look at the inefficient set in terms of technical inefficiency. The dummy variable placed on every coefficient assumes a value of one for the inefficient set and zero otherwise, giving us the unrestricted model. For the first regression the observed $R^2$ value turned out to be 0.303 which tells us that only 30% of the variation in $\theta*$ is explained by the model as follows:

$$\ell n\theta* = \underset{(t=-1.83)}{-0.785} + \underset{(0.108)}{0.003}\ell nx_1 + \underset{(2.371)}{0.098}\ell nx_2$$
$$+ \underset{(0.079)}{0.002}\ell nx_3 - \underset{(-2.125)}{0.097}\ell nx_4 + \underset{(0.395)}{0.005}\ell nx_5; \quad R^2 = 0.303$$

For the second regression, the $R^2$ value becomes 0.939. We now use an F-test to test the hypothesis that there is no explanatory value in these dummy variables and therefore no difference in the input output relationship of the two sets, the inefficient and the efficient. The F-test is of the form

$$F_{m,N-m} = \left[(R_{UR}^2 - R_R^2)/m\right]\left[(1 - R_{UR}^2(N-m)\right]^{-1}$$

$$= \frac{(0.938677 - 0.32976)/5}{(1 - 0.938677)/17} = 35.24 \qquad (5.3.3)$$

Because the computed value of F of 35.24 exceeds the 5% critical value of 2.81, we reject the null hypothesis that there is no difference in technical efficiency between the efficient and the inefficient airlines.

There is an alternative way to analyze the impact of technical efficiency. This is by regressing unit profits ($\pi$) defined by yield minus unit costs on the two measures $\theta*$, $\phi*$ of technical efficiency. The statistical results are as follows:

$$\pi = \underset{(t=-2.41)}{-0.506} + \underset{(2.51)}{0.502}\theta* + \underset{(1.88)}{0.031}\phi* \qquad R^2 = 0.251$$

Following the same dummy variable model as in (5.3.3), we also consider a nested model, which ignores the differences between the technically efficient and inefficient airlines and an F-test is done to test if the inefficient airlines significantly differ from the technically efficient ones. The result is as follows:

$$F_{2,20} = \frac{(0.2996 - 0.2505)/2}{(1 - 0.2996)/(22-2)} = 0.7003$$

The critical 5% value of $F_{2,20} = 3.49$. Hence we cannot reject the null hypothesis that inefficient airlines differ from the technically efficient ones. Note however that both the coefficients of $\theta*$ and $\phi*$ are positive implying a positive contribution to profits, although the TE parameter $\theta*$ is more important and also statistically significant at the 5% level of t-test. However, only about 25% of the variability of profits is explained by the technical efficiency parameters.

Broadly speaking, we may summarize the four overall conclusions reached in this section. First, our DEA results on the overall and allocative efficiency for the world airlines broadly agree with the time series results of stochastic production and cost frontiers estimated by Oum and Yu. Our results however are firm-specific and not based on the assumption of a specific distribution. Secondly, there exist several ways by which the DEA results can be combined with a statistical regression approach. One such important result shows that the volume of passenger demand generates scale economies in the cost function. Thirdly, both the TE and AE measures of efficiency have procyclical variations over the whole period 1980-95 due to the cyclical variations in GDP in different countries, which influence the passenger travel. Finally, the TE measure is only a partial measure, which only explains less than 26% of the variation in gross profits for the world airlines. The OE measure is more meaningful.

## 5.4 Telephone Industry: Efficiency and Profitability

In the field of world communications, telephone industry plays a significant role. Changes in technology in this industry have direct impact on the globalization of trade in recent times and in collaboration with computer technology, the telephone industry is undergoing a process of rapid advancement, which will have a strong impact on the volume and direction of international trade and information technology.

As a benchmark application we consider the data set of NTT (Nippon Telegraph and Telephone) in Japan over the 39-year period 1953-1992, which was originally used by Sueyoshi (1997) to estimate by DEA models several concepts of efficiency, e.g., TE, OE, AE and returns to scale. In this application, Sueyoshi considered each year as a separate DMU, so that different years provide the cost and returns performance of different DMUs or enterprises of the common carrier.

This data set uses three outputs and three inputs. The three output measures are (a) variable charge revenue ($y_1$) , mainly generated from toll telephone services; (b) fixed charge revenues ($y_2$), mainly generated from local telephone services and (c) other miscellaneous revenues. The three input measures are total assets ($x_1$), total employees ($x_2$) and total access lines ($x_3$). For cost frontier estimation by the DEA model total operating expenses (c) are used as a cost measure and input prices associated with the cost are computed by $q_1$ = total depreciation and amortization/total assets, $q_2$ = unit labor costs and $q_3$ = total maintenance and repair cost/total access lines, where $c = q_1x_1 + q_2x_2 + q_3x_3$. These input measures are quite standard, as the international organizations such as OECD (Organization for European Cooperation and Development) have used them to evaluate the performance of the public communications network.

The technical efficiency model uses the formulation (5.1.2) with two boundary constraints on the weights $\lambda_j$ as

$$L \leq \sum_{j=1}^{N} \lambda_j \leq U \qquad (5.4.1)$$

where L and U are the lower and upper bounds respectively, which are pre-assigned on an a priori basis. The cost efficiency model uses the following LP formulation

$$\text{Min} \sum_{i=1}^{m} q_{ki} x_i = c_k$$

$$\text{s.t.} \quad \sum_{j=1}^{N} X_j \lambda_j \leq x; \sum_{j=1}^{N} Y_j \lambda_j \geq Y_k \qquad (5.4.2)$$

$$L \leq \sum_{j=1}^{N} \lambda_j \leq U; \lambda_j \geq 0; x \geq 0; \quad j=1,2,...,N$$

where $q_{ki}$ is the i-th input price or cost incurred by the k-th unit or $DMU_k$. If the input prices are the same, i.e., $q_{ki} = q_i$ for all DMUs and $L = U = 1$, we obtain the standard model (5.4.1) of overall efficiency discussed before. Denote the optimal values of (5.4.2) by asterisks. Then the overall efficiency of $DMU_k$ is

$$OE_k = c_k^* / c_k = q_k' x^* / q_k' X_k$$

and the dual problem reduces to

$$\text{Max } \alpha' Y_k + u_1 L - u_2 U$$
$$\text{s.t.} \quad \alpha' Y_j \leq \beta' X_j + u_1 - u_2; j = 1, 2, ..., N$$
$$\beta_i \leq q_{ki}; 1, 2, ..., m$$
$$\alpha, \beta \geq 0; u_1, u_2 \geq 0$$

Sueyoshi defined the ratio $w = (u_1 L - u_2 U) / c_k^*$ and characterized the various measures of scale efficiency e.g.,

Increasing returns to scale IRTS iff $u_1^* > 0$ and $0 < w < 1$

Constant returns to scale CRTS iff $u_1^* = u_2^* = 0$ and $w = 0$

Decreasing returns to scale DRTS iff $u_2^* > 0$ and $w < 0$

Where iff denotes the term 'if and only if'. Following Baumol, Panzar and Willig (1982) he defined the degree of scale economies (DSE) at $(c_k, Y_k)$ as

$$\text{DSE}_k = c_k^* \left( \sum_{r=1}^{s} \alpha_r^* y_{rk} \right)^{-1} = c_k^* / \alpha^{*\prime} Y_k$$

and showed the equivalence of IRTS with $\text{DSE}_k > 1$, CRTS with $\text{DSE}_k = 1$ and DRTS with $\text{DSE}_k < 1$.

Furthermore the two bounds L and U may pre-assigned in several ways to generate different forms of DEA model discussed in the literature, e.g.,

(i)      $L = 1, U = 1$ yields the BCC model
(ii)     $L = 0, U = \infty$ yields the CCR ratio form
(iii)    $L = 0, U = 1$ yields the hybrid model (IRTS)
(iv)    $L = 1, U = \infty$ yields the hybrid model (DRTS)

The standard model (5.4.1) discussed before follows case (i) above with $L = U = 1$. The empirical results obtained by Sueyoshi (1997) found several interesting aspects of efficiency of the corporate performance of the common carrier NTT. First of all, NTT has exhibited IRTS in the 35 annual time periods (from 1953 to 1987 and 1989-90) out of 39. CRTS (i.e., $u_1^* = 0, u_2^* = 0$) obtained in 4 annual periods from 1987-89 and from 1990-92, when DSE = 1.0. Secondly, the size of DSE has been gradually falling from DSE = 2.56 in 1953-54 to DSE = 1.00 in 1991-92. Thus the scale expansion of NTT operation is not as important today as in the initial stage (1953-60).

If we interpret the case TE = 1.0 = AE where OE = 1.0 as one of equilibrium defined by the tangency of the isoquant and the iso-cost line, then we have 7 out of 39, i.e., 17.9% as equilibrium points. This shows the predominance of disequilibrium points i.e., about 82.1% of the DMUs are not in market equilibrium. In a dynamic context this implies a very significant role of adaptivity and learning over time. This is especially true for those inputs, which are in the form of capital inputs, which have output effects for several years ahead. In this NTT case the access lines can be treated much like durable capital inputs which can be distinguished from the short run inputs such as labor and assets. To see the implication of such capital inputs we formulate an access lines efficiency model as

$$\text{Min } q_3 x_3 + \theta_1 + \theta_2$$

$$\text{s.t.} \quad \sum_{j=1}^{N} x_{3j} \lambda_j \le x_3; \sum_{j=1}^{N} x_{ij} \lambda_j \le \theta_i x_{ik}; \quad i = 1, 2$$

$$\sum_{j=1}^{N} y_{rj} \lambda_j \ge y_{rk}; \quad r = 1, 2, 3; \quad N = 39 \qquad (5.4.3)$$

$$\sum_{j=1}^{N} \lambda_j = 1; \theta_i \ge 0; \lambda_j \ge 0$$

This model is compared with a general input-specific TE model where the objective function of (5.4.3) is replaced by $\text{Min}(\theta_1 + \theta_2 + \theta_3)$ and the first two constraints are modified as

$$\sum_{j=1}^{N} x_{ij}\lambda_j \le \theta_i x_{ik}; \quad i = 1,2,3 \tag{5.4.4}$$

and $\quad \theta_i > 0, \; i = 1,2,3$

Table 5.4.1 shows the DEA results for the two input-specific models (5.4.3) and (5.4.4). Two points emerge very clearly. One is that the difference between the optimal and actual levels of access lines is quite large on an overall basis for the years when $\theta_3^* < 1.0$. For example the NTT had an average slack of excess inputs $(x_3 - x_3^*)$ about 16.3% higher than that needed for production. Secondly, the marginal contribution of access lines on total costs is much higher than the two other inputs. This may be quantitatively indicated by running a nested regression model which regresses total input costs c on the three inputs $x_1, x_2, x_3$, each with a dummy variable indicating if the unit is 100% efficient in the OE or TE measure. The nested model used here is of the linear form

$$c = b_1 + b_2 D + b_3 x_1 + b_4 D x_1 + b_5 x_2 + b_6 D x_2$$
$$+ b_7 x_3 + b_8 D x_3 + \text{error}$$

where $D = 1$ if the observation is TE or OE and zero if it is inefficient. The regression estimates are as follows:

$$\hat{b}_1 = \underset{(t=4.5)}{544.8}; \quad \hat{b}_2 = \underset{(-0.71)}{-484.8}; \quad \hat{b}_3 = \underset{(8.28)}{0.366}; \quad \hat{b}_4 = \underset{(-2.13)}{-0.532}$$
$$\hat{b}_5 = \underset{(-5.97)}{-0.381}; \quad \hat{b}_6 = \underset{(0.640)}{0.257}; \quad \hat{b}_7 = \underset{(1.61)}{0.196}; \quad \hat{b}_8 = \underset{(2.45)}{1.25}$$
$$R^2 = 0.998; \text{ Adjusted } R^2 = 0.992; DW = 1.35$$

These results indicate that for the inefficient years both assets $(x_1)$ and access lines $(x_3)$ have a higher contribution to cost than the employees $(x_2)$, which has a negative coefficient. On the other hand the coefficients for the efficient years indicate that the access lines input have the highest contribution to total costs. The restricted model reinforces these results.

For instance the estimates of the following model

$$c = a_1 + a_2 x_1 + a_3 D x_1 + a_4 x_2 + a_5 D x_3 \tag{5.4.5}$$

yield the following results

$$\hat{a}_1 = 392.28; \quad \hat{a}_2 = 0.435; \quad \hat{a}_3 = -0.476; \quad \hat{a}_4 = -0.297$$
$$\quad (t=4.25) \qquad \quad (61.04) \qquad \quad (-4.44) \qquad \quad (-6.37)$$
$$\hat{a}_5 = 1.188; \quad R^2 = 0.999; \quad \text{Adjusted } R^2 = 0.998$$
$$\quad (4.78)$$

When the above cost function (5.4.5) is run in loglinear form, the coefficient of $Dx_3$ turns out to be 0.971 with a t-value of 6.12, whereas the other two inputs yield coefficients that are not significant at 5% of t-statistics.

Since the data on both input and output prices are available here, one could analyze here the impact of TE and AE on gross profits (p) defined as total revenue minus total input costs. The efficient estimates are then obtained by least squares

$$\pi = -684.77 - 1487.4 \, AE + 2596.3 \, TE \qquad R^2 = 0.431$$
$$\quad (t=-0.52) \qquad (-1.02) \qquad \quad (5.16)$$

$$\pi = -1942.4 + 2415.8 \, TE; \qquad\qquad R^2 = 0.415$$
$$\quad (-4.49) \qquad (5.12)$$

$$\ell n\pi = 1.657 + 0.419 \, \ell n AE + 1.481 \, \ell n TE; \quad R^2 = 0.031$$
$$\quad (6.55) \qquad (0.088) \qquad\qquad (0.969)$$

Clearly the impact of allocative efficiency is not statistically significant; also the multicollinearity between AE and TE is very high. Technical efficiency however has a highly significant impact on profits.

Two other aspects of efficiency can be tested from our estimates of TE and AE from this data set. One concerns the stability of the results on an overall basis for the entire 39-year period. The other is the impact of input price fluctuations on the DEA estimates of OE and AE. For the first case we compare the coefficient of variations (CV) of the observed and optimal inputs. The results are as follows:

|  | CV Estimates | | |
|---|---|---|---|
|  | Assets $(x_1)$ | Employees $(x_2)$ | Access Lines $(x_3)$ |
| Observed | 0.813 | 0.226 | 0.785 |
| Optimal (TE) | 0.867 | 0.245 | 0.839 |
| Optimal (OE) | 0.880 | 0.232 | 0.854 |
| Mean (observed) | 5491.8 | 2575 | 2365 |
| Mean (TE) | 5162.4 | 2349 | 2217 |
| Mean (OE) | 4923.0 | 2238 | 2116 |

Clearly the points on the efficient frontier have higher coefficient of variation. Of the three inputs, access lines that are much like a capital input exhibit the highest increase in CV compared to the observed.

For testing the impact of price fluctuations on overall efficiency we have a choice of two objective functions: either linear or quadratic

(1)      $\text{Min } C = \sum_{i=1}^{3} (\overline{q}_i / \sigma_{ii}) x_i$

(2)      $\text{Min } C = \sum_{i} \overline{q}_i x_i - \alpha \sum_{i=1}^{3} \sum_{j=1}^{3} x_i \sigma_{ij} x_j$          (5.4.6)

subject to the same constraints as

$$\sum_{j=1}^{N} X_j \lambda_j \le x; \sum_{j=1}^{N} Y_j \lambda_j \ge Y_k; \Sigma \lambda_j = 1, \lambda_j \ge 0$$

Here $\sigma_{ij}$ is the covariance of input prices $q_i$ and $q_j$ and the mean levels of prices are denoted as $\overline{q}_i$. The sample estimates of $\overline{q}_i$ and $\sigma_{ij}$ can be obtained from the data. On using these one can apply either the LP model given by the first objective function of (5.4.6) or a quadratic programming (QP) model in the second case. The results are as follows:

|                              | 1953-70 | 1971-92 |
|------------------------------|---------|---------|
| Average Observed Cost        | 444.77  | 4549.45 |
| Average Optimal Cost (LP)    | 342.12  | 4391.15 |
| Average Optimal Cost (QP)    | 320.31  | 4289.24 |

$(\alpha = 0.5)$

Clearly the risk adjusted optimal cost allows for risk adverse efficiency and the quadratic programming model allows for variable substitution rates among the three inputs. Moreover the QP model increases the proportion of OE efficient DMUs from 17.9% to 32.1%. This shows that a nonlinear specification in the DEA model is more flexible in allowing for input substitution.

Broadly speaking we may summarize the three major conclusions for this empirical application. First, the NTT industry has exhibited significant increasing returns to scale in 35 out of 39 year period, although the degree of IRTS has declined in recent years. Second, the access lines input exhibits a more significant contribution to total cost than the other inputs. This implies that long run planning in the optimal utilization of this input is likely to pay off much more than the other inputs. Finally, technical efficiency has a strong and significant positive correlation with gross profits. This implies that any policy measures aimed at improving risk-adjusted TE would improve the average profit prospects.

| Year | Access Lines Efficiency | | $\theta_1^*$ | Input Specific Efficiency | |
| | Optimal Input $(x_3^*)$ | Actual Input $(x_3^*)$ | | $\theta_2^*$ | $\theta_3^*$ |
|---|---|---|---|---|---|
| 1953-54 | 176.9 | 176.9 | 1.0 | 1.0 | 1.0 |
| 1954-55 | 176.9 | 176.9 | 1.0 | 1.0 | 1.0 |
| 1955-56 | 208.6 | 208.6 | 0.963 | 0.996 | 0.963 |
| 1956-57 | 228.5 | 239.7 | 0.994 | 0.964 | 0.966 |
| 1957-58 | 263.8 | 263.8 | 1.0 | 1.0 | 1.0 |
| 1958-59 | 264.3 | 290.3 | 0.963 | 0.931 | 0.928 |
| 1959-60 | 296.3 | 321.6 | 0.968 | 0.911 | 0.937 |
| 1960-61 | 331.9 | 363.3 | 0.897 | 0.891 | 0.927 |
| 1961-62 | 378.1 | 415.3 | 0.854 | 0.871 | 0.922 |
| 1962-63 | 407.5 | 478.1 | 0.791 | 0.835 | 0.862 |
| 1963-64 | 464.9 | 547.7 | 0.772 | 0.803 | 0.858 |
| 1964-65 | 529.0 | 633.9 | 0.764 | 0.774 | 0.842 |
| 1965-66 | 590.1 | 730.3 | 0.741 | 0.748 | 0.814 |
| 1966-67 | 689.6 | 846.6 | 0.761 | 0.735 | 0.820 |
| 1967-68 | 804.4 | 988.9 | 0.782 | 0.721 | 0.818 |
| 1968-69 | 915.5 | 1136.2 | 0.799 | 0.708 | 0.811 |
| 1969-70 | 1052.8 | 1300.5 | 0.822 | 0.699 | 0.813 |
| 1970-71 | 1180.2 | 1517.3 | 0.818 | 0.689 | 0.780 |
| 1971-72 | 1316.6 | 1731.3 | 0.799 | 0.681 | 0.763 |
| 1972-73 | 1518.5 | 2098.5 | 0.789 | 0.687 | 0.725 |
| 1973-74 | 1749.4 | 2416.6 | 0.798 | 0.686 | 0.725 |
| 1974-75 | 1890.5 | 2744.4 | 0.765 | 0.685 | 0.690 |
| 1975-76 | 2111.7 | 3034.3 | 0.771 | 0.688 | 0.697 |
| 1976-77 | 2489.7 | 3242.7 | 0.837 | 0.707 | 0.769 |
| 1977-78 | 3241.3 | 3394.5 | 0.984 | 0.754 | 0.955 |
| 1978-79 | 3398.2 | 3549.4 | 0.958 | 0.756 | 0.958 |
| 1979-80 | 3575.1 | 3704.6 | 0.945 | 0.766 | 0.965 |
| 1980-81 | 3699.4 | 3849.0 | 0.914 | 0.777 | 0.961 |
| 1981-82 | 3806.4 | 3933.1 | 0.895 | 0.781 | 0.968 |
| 1982-83 | 3978.6 | 4110.4 | 0.905 | 0.806 | 0.968 |
| 1983-84 | 4157.6 | 4245.5 | 0.921 | 0.833 | 0.979 |
| 1984-85 | 4288.3 | 4401.9 | 0.917 | 0.840 | 0.975 |
| 1985-86 | 4441.6 | 4486.1 | 0.922 | 0.911 | 0.990 |
| 1986-87 | 4599.5 | 4672.5 | 0.962 | 0.954 | 0.984 |
| 1987-88 | 4797.7 | 4797.7 | 1.0 | 1.0 | 1.0 |
| 1988-89 | 5199.2 | 4990.4 | 1.0 | 1.0 | 1.0 |
| 1989-90 | 5175.0 | 5199.2 | 1.0 | 1.0 | 1.0 |
| 1990-91 | 5408.4 | 5408.4 | 1.0 | 1.0 | 1.0 |
| 1991-92 | 5580.0 | 5580.0 | 1.0 | 1.0 | 1.0 |

Table 5.4.1 Access Lines Efficiency and Input Specific Efficiency

## 5.5 Portfolio Efficiency Frontier: Risk and Efficiency

Evaluation of investment portfolios under uncertain conditions of the stock market has been an important area of applied research in financial economics. Recent volatility in the stock market has generated more intense interest today. Mutual funds investment in U.S. have grown tremendously over the last decade and their performance has been frequently compared with the overall market index like S&P 500. Our object in this section is to evaluate the relative efficiency of performance of 74 mutual fund portfolios over eleven years (1988-98). The annual data set is from Morningstar, which also provides estimates of Jensen's alpha ($\alpha$) and beta ($\beta$) of the CAPM (Capital Asset Pricing Model), the estimates of mean return, its variance and the Sharpe index (ratio).

A set of 60 funds out of 74 is selected in our study, divided into four groups with 15 in each group. The four groups are: (a) growth funds which emphasize capital growth in their choice of stocks, (b) balanced funds, which place more weight on risk minimization through diversification of stocks, (c) income funds which emphasize continued income safety and (d) technology and communication funds, which play a major role in NASDAQ composite index. This classification of mutual funds follows the codes on portfolio objectives listed by Weisenberger Financial Services for all funds traded in the NYSE market.

Our empirical analysis applies the DEA model to measure the relative efficiency of mutual fund portfolios mentioned above. This nonparametric method of measuring portfolio performance is better than the two other measures often used in financial economics e.g., Jensen's alpha and the Sharpe index defined as the ratio of excess return on a portfolio to its standard deviation. The reasons are several: no need for a benchmark portfolio, no need to exclude transactions costs and no need to assume the normality of return distribution. Moreover the DEA model permits the calculation of an efficiency index for each mutual fund, whereas the Markowitz-Tobin method of estimating a parametric mean variance efficiency frontier provides only statistical averages. Recently Murthi, Choi and Desai (1997) applied the DEA method to a total of 731 mutual funds divided into seven categories, e.g., aggressive growth, asset allocation, equity income, growth, growth-income, balanced and income for the third quarter of 1993. Besides calculating the DEA efficiency measures they performed a regression analysis to test for the source of variation in mean efficiency scores across the different categories. The three broad conclusions reached by their study are as follows. First, a striking result is that along the DEA efficiency frontier the risk measured by standard deviation has virtually no slacks throughout all investment categories. This means that the mutual funds considered here are all approximately mean-variance efficient. Second, the mean efficiency scores are found to be unrelated to mean expense ratios, mean loads or mean turnover, e.g.,

| Intercept | Expense Ratio | Loads | Turnover | Standard Deviation | $R^2$ |
|-----------|---------------|---------|----------|--------------------|-------|
| 0.790 | 0.090 | -0.0007 | -0.0001 | -0.0145 | 0.09 |
| (t=10.8) | (1.39) | (-0.03) | (-0.38) | (-1.00) | |

Note that all the coefficients are not statistically significant from zero. This implies that funds that charge a higher transaction costs in the form of expenses and fees do not more than compensate for the greater cost in terms of efficiency. This finding is contrary to the efficiency hypothesis of Grossman and Stiglitz (1980) who argue that in an efficient market the informed investors should be compensated with higher returns for their investment than the uninformed investors. Finally, the efficiency is not related to the size of the funds measured by the mean asset value. This follows from the fact that the correlation between size and efficiency scores is not significantly different from zero in a statistical sense for the entire data set.

Our empirical application has several new features compared to the study by Murthi et al (1997). First of all, we use more directly two outputs and five inputs, whereas their study used a ratio model to maximize the ratio of return to weighted input costs for the reference fund. The five inputs are the costs of mutual funds such as loads, turnover, expense ratio, standard deviation of return and the correlation between the fund and the market return as S&P 500. The last input exhibits market risks in terms of the divergence from the efficient market as viewed in CAPM (capital asset pricing model). Our two outputs are annual returns and a proxy represented by the return skewness. In earlier studies Sengupta (1989, 1996c) found that skewness of returns has a significantly favorable impact on mean returns in bullish markets and since the eleven year period (1988-98) has exhibited an optimistic view of the overall market, we considered skewness of return as a second output in our formulation. Second, we set up the TE model (5.1.1) as before with five inputs and two outputs for the four groups of mutual funds, with the technology and communication group as the new technology-intensive funds, which showed phenomenal growth along with the Internet and Personal Computer industries. We use the DEA results on the efficiency scores ($\theta$) for different funds. Third, the linear regression method is employed to explain the variations in TE scores $\theta_j^*$ in terms of loads, risk, expense ratio, turnover, return and skewness. We use two measures of risk: standard deviation of return and beta. A nested model is used which incorporates a dummy variable for the inefficient and the efficient funds.

Finally, the mean variance efficiency frontier is estimated from the following DEA model

$$\text{Max } \phi$$

$$\text{s.t.} \quad \sum_{j=1}^{N} \mu_j \lambda_j \geq \phi \mu_k ; \sum_{j=1}^{N} \mu_j^2 \lambda_j \geq \mu_k^2 \qquad (5.5.1)$$

$$\sum_j \sigma_j^2 \lambda_j \le \sigma_k^2 ; \Sigma \lambda_j = 1, \lambda_j \ge 0$$

On using the Lagrange multipliers $a_0$, a, b, $b_0$ the variance frontier becomes

$$\sigma_j^2 \le (1/b)\left[a\mu_j + a_0\mu_j^2 + b_0\right] \qquad (5.5.2)$$

where $\mu_j$ and $\sigma_j^2$ are the mean and variance of returns of mutual fund j=1,2,...,60 and the multiplier a is free in sign. Assuming equality in (5.5.2) the variance frontier finally reduces to

$$\sigma_j^2 = a_1 + a_2\mu_j + a_3\mu_j^2 \qquad (5.5.3)$$

where $a_1 = b_0/b$, $a_2 = a/b$ and $a_3 = a_0/b$. Note that this frontier equation can be directly compared with the corresponding regression estimate.

The empirical results on DEA efficiency are reported in Table 5.5.1, which shows that the growth and technology funds are more efficient than the income and balanced funds. For instance 87% of growth funds are efficient, whereas only 60% of the income funds are efficient. About 72% of the funds are efficient on an overall basis. The average slacks for four groups of funds appear as follows:

| Fund | Load | Std. Dev | Exp. Ratio | Turn-over | Cov w/ market | Mean | Skewness |
|------|------|------|------|------|------|------|------|
| Balanced | 0.0 | 0.037 | 0.256 | 39.76 | 3.46 | 0 | 0 |
| Growth | 1.58 | 0.0 | 0.24 | 0.0 | 13.05 | 0.41 | 0 |
| Tech & Comm | 0.27 | 0.86 | 0.42 | 47.13 | 8.04 | 0.55 | 0.69 |
| Income | 0.32 | 0 | 0.18 | 19.36 | 9.74 | 0.09 | 0.07 |

This shows that there is no significant slack in standard deviation and skewness for the balanced, growth and income funds. However there is significant slack in covariance with market and turnover expense. Since turnover is an activity involving higher uncertainty one would expect that it will be a major source of inefficiency. However, the growth fund managers are turnover efficient on the average, although their expense ratio is lower than that of technology and balanced funds. On an overall basis our results tend to support the broad conclusion reached by Ippolito (1989) in his regression estimate that the extra costs due to information gathering would equal extra returns, so that the market would be efficient in the CAPM model in terms of net adjusted return.

The regression of our efficiency measure $(\theta^*)$ on the various inputs produces the following estimates

| Balanced Funds | θ | Technology & Communication | θ |
|---|---|---|---|
| Federal Stock & Bond A | 1 | Alliance Technology A | 1 |
| Evergreen Blncd B | 0.87 | Fidility Select Computers | 1 |
| Fedility Adviser Blncd T | 1 | Fidility Select Electronics | 1 |
| Fedility Blncd | 0.97 | Fidility Select Software & Comm | 1 |
| Fedility Puritan | 1 | Fidility Select Technology | 0.9 |
| Founders Blncd | 0.88 | Franklin Dyna Tech A | 1 |
| Franklin Income A | 1 | John Hancock Global Tech'A | 0.9 |
| Greenspring | 1 | Invesco Technology II | 1 |
| Hotchkis & Wiley blncd | 1 | Kemper Technology A | 1 |
| IDS Mutual A | 1 | T. Rowe Price Science & Tech | 1 |
| Kemper Total Return A | 0.89 | Seligman Commun & Info A | 1 |
| Linder Dividend Inv | 1 | United Science & Tech A | 1 |
| MainStay total Return B | 1 | Fidility Select Multimedia | 0.9 |
| Merrill Lynch Capital A | 1 | Fidility Select Telecomm | 0.9 |
| MSF Total Return A | 0.95 | Flag Investors Commun A | 1 |

| Growth Funds | θ | Income Funds | θ |
|---|---|---|---|
| Brandywine | 1 | Alliance Muni Income CA A | 0.8 |
| Columbia Special | 1 | American Century CA Long-Term | 1 |
| Delaware DelCap A | 1 | American Century CA High-Yield | 1 |
| Evergreen Aggressive Growth A | 1 | California Investment Tax-Free | 0.8 |
| Federated Growth Strategies A | 1 | Dreyfus CA Tax-Exempt Bond | 1 |
| Fedility OTC | 1 | Fidelity Spartan CA Muni Income | 0.9 |
| Fortis Growth A | 1 | Franklin CA Insurance Tax-Free | 1 |
| Founders Special | 1 | Franklin CA Tax-Free Income A | 1 |
| IDS Strategy Aggressive B | 0.98 | Merrill Lynch CA Muni Bond B | 0.8 |
| Invesco Dynamics | 1 | T. Rowe Price CA Tax-Free bond | 1 |
| MFS Emerging Growth B | 1 | Pantum CA Tax-Examot Inc A | 0.9 |
| Neuberger Berman Manhattan | 1 | Scudder CA Tax-Free | 0.9 |
| Nicholas-Applegate Grwth Eqty A | 0.84 | Smith Barney CA Municipals A | 1 |
| PainWebbe Growth A | 1 | Vanguard CA Insured Long | 1 |
| Parkstone Mid Capitilization Instl | 1 | Tax-Exempt Fund of CA | 1 |

Table 5.5.1  DEA Efficiency (θ)

| Funds | Intercept | Load | Std dev | Exp ratio | Turnover | Return | Skewness | $R^2$ |
|-------|-----------|------|---------|-----------|----------|--------|----------|-------|
| Inefficient | 0.86 | -0.01 | -0.02 | -0.04 | -0.0 | 0.03 | 0.04 | 0.91 |
| | (t=36.8) | (-2.4) | (-4.08) | (-2.7) | (-3.9) | (5.6) | (4.1) | |
| Efficient | 1.00 | 0.0 | -0.0 | 0.0 | 0.0 | 0.0 | 0.0 | 0.81 |
| | (4.99) | (2.15) | (3.76) | (2.01) | (3.58) | (5.3) | (-3.5) | |

When we use the data on market beta as a risk measure instead of the standard deviation, the results are broadly similar. Thus our results broadly agree with the argument by Grossman and Stiglitz (1980) that informed investors should be compensated with higher returns for their investment than the uninformed investors, although there are differences between different groups of mutual funds.

The DEA results obtained from the dual model (5.5.3) are also used to estimate the quadratic mean variance frontiers on using a regression method. The overall results are as follows:

| Fund | Intercept | $\mu$ | $\mu^2$ | $R^2$ |
|------|-----------|-------|---------|-------|
| Balanced | 338.1 | -22.68 | 0.58 | 0.32 |
| | (t=1.07) | (-0.47) | (0.44) | |
| Growth | 634.7 | -20.13 | 0.50 | 0.25 |
| | (0.68) | (-0.37) | (0.34) | |
| Technology | 724.0 | -18.87 | 0.27 | 0.21 |
| | (0.74) | (-0.33) | (0.44) | |
| Income | 492.3 | -94.2 | 5.03 | 0.27 |
| | (1.02) | (-0.75) | (0.64) | |
| S&P 500 | 344.3 | -6.88 | 0.328 | 0.31 |
| (Observed data) | (5.08) | (-2.23) | (3.51) | |

Although the regression coefficients are not statistically significant, the results uniformly indicate a negative coefficient for mean return, while the coefficient of $\mu^2$ is positive all throughout. Thus the variance frontier is a strictly convex function of mean return where

$$\frac{\partial \sigma^2}{\partial \mu} \begin{cases} < 0, \text{if } a_3\mu < |a_2| \\ > 0, \text{otherwise} \end{cases}$$

This type of asymmetry has been found in earlier studies by Sengupta (1996), who noted that in bullish markets past trends in variance tend to underestimate future variances. Clearly our DEA results bear out this asymmetry hypothesis.

To summarize our broad results in this section we mention two hypotheses. First, the relative efficiency evaluation of different groups of mutual funds by the DEA approach without using any benchmark for comparison shows that in bullish markets some funds may outperform the overall market with a high probability. Secondly, the DEA results on the mean variance frontier imply a certain degree of asymmetry in the impact of mean on variance, positive in a bearish market but negative in a bullish market. Indirectly it casts a serious doubt on the normality of return distribution.

# 6. Economic Theory and DEA

The economist views DEA as a partial equilibrium characterization of efficiency of firms within an industry. In case of technical or production efficiency the market prices are not used at all. True, in allocative efficiency the prices are used but the role of market demand is rarely incorporated in any detail. The economist's outlook on partial equilibrium views it as a special case of general equilibrium and hence efficiency should be viewed at three complementary levels: the producers, the consumers and the agents (traders) who coordinate production and consumption. The DEA models concentrate on production side only and that too at a given industry level. Farrell who first proposed a radial measure of technical efficiency in terms of a conical hull also proposed a second concept for inter-industry comparison. This concept is called *structural efficiency*, which broadly measures the degree to which an industry keeps up with the performance of its own best practice firms. Farrell did not quantify this industry level measure, although he mentioned the possibility of using market prices in some sense to characterize the industry level efficiency. Thus it is a measure at the industry level of the extent to which its firms are of optimum size, to which its high-cost firms are squeezed out or reorganized, to which the industry production level is optimally allocated the firms in the short run. Clearly the Leontief-type input-output model provides such a framework of inter-industry or inter-sectoral comparison, where each industry may represent a sector, which has interdependence with other sectors.

Besides the general equilibrium viewed of economic efficiency, the notion of disequilibrium analysis is emphasized in economic theory of partial equilibrium. The Walrasian adjustment process assumes that excess demand (supply) tends to increase (decrease) equilibrium prices in competitive markets and under normal conditions this adjustment process converges. However cases of market failure may occur and disequilibrium may persist. The economics of market disequilibrium has been analyzed in economic literature quite extensively both theoretically and statistically. Benassy (1982) has analyzed the implications of market disequilibrium in terms of rationing, spillover effects and the stochastic aspects of demand. Maddala (1983) has discussed the econometric problems of estimation in such frameworks. Clearly there is a need for DEA theory to incorporate such problems of disequilibrium.

The third economic aspect under emphasized in DEA theory is the role of stochastic factors in the production and allocative decisions. The theory of efficiency evaluation under risk aversion or uncertainty needs to be incorporated into DEA theory both analytically and theoretically. Analytically one needs to apply DEA models to investment markets, where financial economics usually considers the capital asset pricing models based on mean variance efficiency frontier. Theoretically one may need to introduce specific distributional assumptions behind the input-output data and their prices and characterize the stochastic production or cost frontier.

Finally, the allocative efficiency model of DEA has great scope of application in optimal production planning problems in the short run. Sequential methods of estimation and control, which are frequently applied in modern optimal control theory, are most suitable in this framework. Methods of learning and adaptivity are of great use here, when the environment involves stochastic factors.

This chapter discusses some of these economic issues with a view that it provides a bridge between DEA and economic theory.

## 6.1 Economics of Disequilibrium

There are two basic sources of disequilibrium in a DEA model. One is due to the fact that the total supply of output determined by the optimal DEA model may not be equal to the total demand for the whole industry. In this case realized supply (demand) would differ from the perceived supply (demand) and market adjustments through price or quantity changes will be needed to ensure equilibrium. Benassy (1982) has discussed various types of adjustment schemes, both Walrasian and non-Walrasian, which would yield equilibrium. The case of stochastic demand and expectations are especially important in this framework. The second source of disequilibrium in DEA model is due to the fact that it is based essentially on a pseudo-production function, rather than a real production function. For a real production function one needs the input data in the form $(x_{irj})$, where $x_{irj}$ denotes input i allocated to product r for the firm j. But the DEA model uses only the aggregate data $x_{ij}$, ignoring the output specific allocation of each input i. Thus consider the production function (5.1.7) in Chapter 5

$$\alpha' Y_j \le \beta' X_j - \alpha_0; \quad j=1,2,\ldots,N$$

If $DMU_j$ or firm j is technically efficient in terms of the DEA model then it follows

$$\alpha'^* Y_j = \beta'^* X_j - \alpha_0^* \tag{6.1.1}$$

Note that the terms $\alpha'^* Y_j$ and $\beta'^* X_j$ are weighted outputs and inputs, where input allocation to specific outputs is completely ignored (see e.g., Sengupta (1996c,d)).

Consider now an extended DEA model which allows output-specific allocations of the input vector $X_j$ for firm j. The firm's objective here is to determine optimal input allocations so as to minimize overall costs:

$$\text{Min } C = \sum_{i=1}^{m} \sum_{r=1}^{s} q_i x_{ir}$$

$$\text{s.t.} \quad \sum_{j=1}^{N} x_{irj} \lambda_j \leq x_{ir}; \quad i \in I_m, r \in I_s \tag{6.1.2}$$

$$\sum_{j=1}^{N} y_{rj} \lambda_j \geq y_{rk}; \Sigma \lambda_j = 1; \lambda_j \geq 0; j \in I_N$$

Here $I_n = \{1,2,...,n\}$ is an index set and $q_i$ denotes the input prices. Denote optimal values by asterisks. The optimal values $x_{ir}^*, \lambda_j^*$ now characterize the production frontier as

$$\sum_{r=1}^{s} \alpha_r^* y_{rk} = \sum_{i=1}^{m} \sum_{r=1}^{s} \beta_{ir}^* x_{irk} - \alpha_0^* \tag{6.1.3}$$

if DMU$_k$ is overall cost-efficient. On comparing (6.1.3) with (6.1.1.) one notes that we have here output-specific marginal productivity parameters $\beta_{ir}^*$ and the actual values $x_{irj}$ of output-specific inputs may be compared with the optimal levels $x_{ir}^*$, so that input inefficiency can be measured by their gap which may be positive or negative.

The demand supply imbalance in market disequilibrium may be illustrated very simply by an output-oriented model with one output and m inputs. In each period t the firm faces the competitive input and output markets, where the input price vector is q, the output price is p and the firm has a supply $\tilde{s}_{jt}$. If $I_t$ denotes inventories at the outset of period t, then the supply of output is given by

$$\tilde{s}_{jt} \leq y_{jt} + I_{jt}; \quad j \in I_N \tag{6.1.4}$$

Let $d_{jt}$ be the demand facing the firm. The the realized sales $s_{jt} = \min(\tilde{s}_{jt}, d_{jt})$ is the minimum of supply and demand. If some unsold goods are left, they are stored and they depreciate at a given rate h

$$I_{j,t+1} = h_j(I_{jt} + y_{jt} - s_{jt}), \quad 0 \leq h_j \leq 1 \tag{6.1.5}$$

The allocative efficiency model under demand uncertainty in DEA framework poses the following decision problem for the firm: how to choose the strategy $(x_t, \tilde{s}_t)$ which maximizes the sum of expected discounted profits

$$\text{Max } E\left[\sum_{t=0}^{\infty} \delta^t (ps_t - \sum_{i=1}^{m} q_i x_{it}), \quad 0 \leq \delta \leq 1\right.$$
$$\text{s.t. the conditions (6.1.4), (6.1.5)}$$

and

$$\sum_{j=1}^{N} x_{ijt} \lambda_{jt} \leq x_{it}; \sum_{j=1}^{N} \tilde{s}_{jt} \lambda_{jt} \geq \tilde{s}_t \qquad (6.1.6)$$

$$\sum_{j=1}^{N} \lambda_{jt} = 1; \lambda_{jt} \geq 0; i \in I_m; j \in I_N$$

Here uncertain demand is in the form of uncertainty in expectations, which causes the imbalance of realized demand from anticipated demand. The inventory variable thus plays a key role in the firm's otpimal production strategy. This model assumes that prices p, q are unchanged in subsequent periods and hence realized sales in period t will be given by $s_t = \min(\tilde{s}_t + I_t, d_t)$. In a more complex model there may be price uncertainty also, thus necessitating the need for speculative inventories.

With several outputs this type of model can be easily transformed. Consider for example a profit-frontier model

$$\text{Max } \pi = p'y - q'x$$
$$\text{s.t.} \quad \sum_{j=1}^{N} X_j \lambda_j \leq x; \sum_{j=1}^{N} Y_j \lambda_j \geq y$$
$$\sum_{j=1}^{N} \lambda_j = 1; \lambda_j \geq 0$$

Here (x,y) denote the unknown input and output vectors to be determined as optimal strategies and $(X_k, Y_k)$ denote their observed levels for DMU$_k$. Two special cases of this model (6.1.6) are important. One is the simpler output-oriented model where demand $(d_r)$ for output $(y_r)$ is subject to a probability distribution $F(d_r)$ and the objective function is to maximize the expected value of total revenue minus expected inventory cost. This yields the model

$$\text{Max } E\left[\sum_{r=1}^{s} p_r \min(y_r, d_r) - \sum_{r=1}^{s} h_r (y_r - d_r)\right] \qquad (6.1.7)$$

s.t.     $X\lambda \leq X_k ; X\lambda \geq y, \lambda'e = 1, \lambda \geq 0$

where $\lambda$ and $y$ are the unknown vectors to be optimally solved for and $h_r$ is the obeserved unit cost of positive inventory for $y_r > d_r$. Denoting optimal values by asterisks, the efficient $DMU_k$ would then satisfy the following marginal condition:

$$F(y_r^*) = (p_r + h_r)^{-1} (p_r - \alpha_r^*)$$
$$\alpha^{*\prime} Y_k = \beta^{*\prime} X_k - \alpha_0^*$$

Clearly higher output price and lower inventory costs would increase the optimal output levels $y_r^*$ which may be compared with the observed outputs $y_{rk}$ in output vector $Y_k$.

The second case is an input-oriented model, where the input decision $x_i$ are equal to planned values $\overline{x}_i$ plus and error term $\varepsilon_l$ with a zero mean and fixed variance. The errors are disturbances such as mistakes or expected difficulties in implementing a planned value $\overline{x}_i$. The planned values $\overline{x}_i$ are the decision variables which have to be optimally chosen by each DMU and the error process $\varepsilon_l$ is realized after the planned value of $x_i(t)$ is optimally selected. The input constraints now turn out to be chance constrained

$$\mathrm{Prob}\left[ \sum_{j=1}^{N} x_{ij}\lambda_{ij} \leq \overline{x}_i + \varepsilon_i \right] = \phi_i, \quad 0 < \phi_i < 1$$

where $\phi_i$ is the tolerance level of the i-th input constraint. The simpler model then takes the following form:

$$\mathrm{Min} \sum_{i=1}^{m} q_i \overline{x}_i$$

s.t.     $$\sum_{j=1}^{N} x_{ij}\lambda_j = \overline{x}_i + w_i ; w_i = F^{-1}(1 - \phi_i)$$

$$\sum_{j=1}^{N} y_{rj}\lambda_j \geq y_{rk} ; \lambda'e = 1, \lambda \geq 0$$

$$I = 1,2,\ldots,m; \ r = 1,2,\ldots,s$$

Clearly the input uncertainty is here capture by the term $w_i$ which depends on the level $\phi_l$ of change constraint, e.g., the higher the level of $w_i$, the lower would be the optimal planned inputs $\overline{x}_i^*$.

The main implication of the general model (6.1.7) is that the input and output gaps measured by $|x_i^* - x_{ik}|, |y_r^* - y_{rk}|$ can be quantified as a source of inefficiency. Even if the two constraints $x \leq X_k$, $y \geq Y_k$ are dropped, we would have the dual model

$$\text{Min } \alpha_0 \quad \text{s.t.} \quad p \leq \alpha, q \geq \beta$$

and
$$\beta'X_j \geq \alpha'Y_j + \alpha_0, j = 1, 2, ..., N$$

$$\alpha, \beta \geq 0, \alpha_0 \text{ free in sign}.$$

Note that a positive (negative or zero) value of $\alpha_0$ indicates the size of increasing (decreasing or constant) returns to scale; hence the dual model can quantify the source of inefficiency of a relatively inefficient DMU in terms of returns to scale. Furthermore if $\alpha^* = p$ and $\beta^* = q$ then the DEA based profit measure $\pi_j^*$ for $DMU_j$ becomes

$$\pi_j^* = p'Y_j - q'X_j + \alpha_0^* \leq 0, j = 1, 2, ..., N$$

We consider now the role of market competition in the efficiency framework. Hence we assume that each firm or $DMU_j$ produces a single homogeneous output denoted by $y_j$, where the total industry output is denoted by $y_T = \sum_{j=1}^{N} y_j$. If N is large and the firms or DMUs are competitive, then the output price p is a constant, unaffected by the size of each individual firm. In this case the price can be viewed as $p = \overline{p} + \varepsilon$ made up of two components: the expected price $\overline{p}$ and a random part $\varepsilon$ with a zero mean and a constant variance $\sigma_\varepsilon^2$. The total cost of inputs for each firm may now be related to output as

$$c(y_j) = c y_j + F_j$$

assuming a linear form, where $F_j$ is the fixed cost and c is marginal cost that is assumed to be identical for each firm. Maximization of expected profits would then yield the LP model:

$$\text{Max } \overline{\pi} = (\overline{p} - c)y - F$$

$$\text{s.t.} \quad \sum_{j=1}^{N} X_j\lambda_j \leq X_k; \sum_{j=1}^{N} y_j\lambda_j \geq y; \lambda'e = 1, \lambda \geq 0 \qquad (6.1.8)$$

where y is the unknown decision variable to be optimally selected. In case the market is imperfectly competitive, the price variable then depends on the output supply of different firms. In the homogeneous output case the firms are all alike, and the inverted demand function can be written as:

$$\bar{p} = a - bY_T, \quad Y_T = \sum_{j=1}^{N} y_j; \quad y_k = y$$

The LP model (6.1.8) would then yield the following optimality conditions:

$$(a\!-\!c)\!-\!b\, Y_T\!-\!b\, y^*\!-\!a^* \leq 0$$

$$\alpha^* y_j - \beta^{*\prime} X_j - \alpha_0^* \leq 0 \qquad\qquad (6.1.9)$$

$$\alpha^*, \beta^* \geq 0, \ \alpha_0^* \ \text{free in sign}$$

If firm k is efficient, then one must have

$$y^* = (a_1/b) - Y_T - \frac{\alpha^*}{b}; \quad y^* > 0$$

$$a_1 = a\!-\!c > 0$$

where $y^* = y_k^*$ is the efficient output of the firm k. If all firms are efficient then $Y_T^* = \sum_{j=1}^{N} y_j^*$ and one obtains

$$Y_T^* = (N/b)(1 + N)^{-1}[a_1 - \alpha^*] \qquad\qquad (6.1.10)$$

Now we introduce organization slack denoted by $s_k$ in the cost function

$$c(y) = (c + s_k)y + F_k; \quad s_k \geq 0$$

Recently Selten (1986) has interpreted this slack concept due to Leibenstein's X-efficiency as a part of the cost function and introduced a 'strong-slack' hypothesis which maintains that this type of slack has a tendency to increase so long as profits are positive, i.e., this slack can be reduced only under the threat of losses. Including this slack-ridden cost into the profit maximization model would yield the optimality condition for the efficient output as:

$$y^* = (a_1 - \alpha - s_k)/b - Y_T; \quad y^* > 0$$

where $y^* = y_k^*$ is the efficient output of the firm k. If all firms are efficient, then

$$Y_T^* = (N/b)(1+N)^{-1}[a_1 - \bar{s} - \alpha^*]$$

$$\bar{p} = a + N(N+1)^{-1}(\alpha^* + \bar{s} - a_1) \qquad\qquad (6.1.11)$$

$$\pi_k^* = (\bar{p} - c s_k)y^* - F_k$$

when $\bar{s} = \sum\limits_{k=1}^{N} s_k/N$ is the average rate of slack. Several implications follow from this set (6.1.11) of efficiency conditions. First of all, the long run pressure of competition would tend to lead to zero profits $\pi_k^* = 0$ for all k=1,2,...,N according to the strong slack hypothesis. In this case the expected price becomes $\bar{p} = c + \bar{s} + (F/y^*)$. This shows that fixed costs have a strong positive role in determining the long run equilibrium price. The higher the average slack rate $\bar{s}$, the higher is the equilibrium expected price. Secondly, as the number N of firms increases, it increases the volume of total industry output $Y_T^*$ and reduces the average price. Finally, as the average slack rate $\bar{s}$ rises (falls), it increases (decreases) the equilibrium price. Note that in case of weak slack hypothesis all profits are not squeezed out and there remains a divergence of individual ($s_k$) from the average slack rate ($\bar{s}$), when the latter is positive. Thus some inefficiency may persist due to the existence of a positive slack.

So far we have assumed tha the expected price $\bar{p}$ is the market clearing price equating market demand and supply. If however this is not the case, then the suply y would differ from demand d, where demand is subject to random fluctuations around the mean level $\bar{d}$. In this framework we have to add to the cost function the costs of inventory and shortage C(y-d). Assuming this cost to be quadratic one may then formalize the decision model:

$$\text{Max } \bar{\pi} = (\bar{p} - c - s_k)y - (1/2)\gamma E(y-d)^2 - F \qquad (6.1.12)$$
$$\text{s.t.} \qquad \text{the same constraint as in (6.1.8)}$$

In this case the optimality conditions for the efficient output becomes

$$y^* = (b+\gamma)^{-1}[(a_1 - \alpha^* - s_k) + \gamma\bar{d} - bY_T] \qquad (6.1.13)$$

where $\gamma$ is the unit cost of inventory or shortage and $\bar{d}$ is the expected level of demand. In this case the marginal impact $\partial y^*/\partial\gamma$ of inventory/excess costs may be either negative or positive according as

$$b(Y_T + \bar{d}) > \text{or}, < (a_1 - \alpha^* - s_k)$$

Again this explains the persistence of some inefficiency, when demand is uncertain and the firm chooses its optimal output by the quadratic criterion of adjusted profits. Furthermore the higher the mean demand, the greater the otpimal level of efficient output y*. In case of perfect competition with each firm a price taker, the optimality condition (6.1.9) reduces to

$$y^* = \overline{d} + (1/\gamma)(\overline{p} - c - s_k - \alpha^*) \qquad (6.1.14)$$

which shows unequivocally that higher inventory costs ($\gamma$) lead ot lower optimal output. Again by comparing the observed output $y_k$ with the otpimal output y*, one could evauate the impact of inefficiency. Note that we still have the comparative static results: $\partial y^* / \partial \overline{p} > 0$ and $\partial y^* / \partial s_k < 0$. Since $\overline{p} = \gamma(y^* - \overline{d}) + c + s_k + \alpha^*$, we have the results:

$$\overline{p} > MC_T, \text{ if } y^* > \overline{d}$$

and

$$\overline{p} < MC_T, \text{ if } y^* < \overline{d} \qquad (6.1.15)$$

where $MC_T = c + s_k + a^*$ is total marginal cost with three components: production cost (c), cost of slack ($s_k$) and the cost of discreipancy of observed output ($\alpha^*$). Clearly the case of multiple output can be handled in a symemtrical way.


## 6.2 Cost Uncertainty and Capacity Utilization

Capacity utilization has two basic roles in industrial price and output policies. The first is one of the basic propositions in macroeconomics, which says that price inflation accelerates as capacity and resource utilization moves higher. The second is the intertemporal implication of changes in capacity inputs, which affect both the fixed and variable costs in the short run. Since every short run production and cost function is conditional on a fixed supply of capacity inputs, the short run cost minimization model may not ordinarily yield the long run cost frontier. We consider here first, a two-period model of capacity expansion and derive the implication of varying the capacity utilization rates. In the next section the long run implication of optimal capacity expansion and its impact on efficient outputs and prices is investigated in some detail.

The term 'capacity' is often viewed as a ceiling on production, or output that is commonly referred to as the engineering definition of capacity. This definition is quite different from the economic view, where full capacity describes a firm's planned or intended level of utilization; the level that reflects satisfied expectations and is built into the capital stock and embodied in the normal working schedule. Two empirical measures of capacity are commonly used in applied work in manufacturing industries. One is the US Federal Reserve Board (FRB) series on capacity indexes which attempt to capture the concept of sustainable practical capacity, which is defined as the greatest level of output that a plant can maintain within the framework of a realistic work schedule, taking into account of normal downtime and assuming sufficient availability of inputs to operate the machinery and equipment in place. Hence this level of output does not necessarily represent either the maximum that can be extracted from the fixed plant (as indicated by utilization rates that sometimes exceed 100 per cent) or the level associated with the minimum point of the short run average cost curve. More specifically, the first step in estimating capacity indexes is to divide an industrial production index ($Q_t$) by a utilization rate ($CU_t$) provided by the Census Department's Survey of Plant Capacity Utilization. This yields an initial estimate of implied capacity: $IC_t = Q_t/CU_t$. However the survey is conducted every four years and firms are asked to report utilization in the fourth quarter of that year. This generally leads to cyclical variability in implicit capacity. To eliminate this cyclical volatility the second step is used to regress implied capacity $IC_t$ on capital stock ($K_t$) and a deterministic function of time as

$$\ln IC_t = \ln K_t + \alpha + \sum_{i=1}^{\tau} \beta_i f_i(t) \qquad (6.2.1)$$

where $K_t$ is the year end capital stock and $f_i(t)$ is an i-th order polynomial defined on time t. the fitted values from these regressions provide baslines for the annual FRB estimates of productive capacity ($C_t$).

A second method in estimating productoin capacity is to use a filter due to Hodrick and Prescott (i.e., HP filter), which decomposes a time series into a permanent and a transitory component. Hodrick and Prescott (1993) define the permanent component as including those variations which are sufficiently smooth to be consistent with slowly changing demographic and technological factors and the accumulation of capital stocks. The HP permanent component is used as a measure of capacity. Then the capacity utilization is calculated as production ($Q_t$) divided by the HP permanent component. In the short run, both demand and supply shocks may cause the deviations of actual output from the permanent component. Note that firms may have several options in regard to raising output above its potential level, e.g., by adding shifts, varying the production line speeds, altering the product mix or even bringing mothballed facilities back into use.

Recently Kennedy (1995) used quarterly data (1960I-1992IV) for U.S. manufacturing to regress the rate of producer price index (PPI) on both utilization rates of FRB and HP capacity variables and found the HP variable to be dominant.

In our approach we combine the two methods above to define a series of capacity levels $CAP_{jt}$ for j-th unit at time t. This is based on two steps. In the first step we assume an additive decomposition of implied capacity into a permanent component ($IC_{jt}^P$) and a transitory component ($\zeta_{jt}$):

$$IC_{jt} = IC_{jt}^P + \zeta_{jt}$$

A filtering method (e.g., Kalman filter ) is applied here to estimate the permanent component, until the random component $\zeta_{jt}$ turns out to be a white noise process. In the second step we use the data on capital stock ($K_{jt}$) and the time variable to regress $IC_{jt}^P$ on $K_{jt}$ and $f_i(t)$ as defined in (6.2.1):

$$\ln \hat{IC}_t = \ln K_t + \alpha + \sum_{i=1}^{\tau} \beta_i f_{ij}(t)$$

Taking antilogs of the dependent variable we obtain the estimate $\hat{CAP}_{jt}$ of capacity. On using this capacity series $z_{jt} = \hat{CAP}_{jt}$ we set up two overall cost minimization models in the DEA frameowrk: one involving the optimal utilization rate $\psi^*$ and the other the optimal capacity $z^*$ and optimal variable inputs $x^*$.

$$\text{Min } \theta + \psi$$

$$\text{s.t.} \quad \sum_{j=1}^{N} \lambda_j x_j \leq \theta X_k ; \sum_{j=1}^{N} \mu_j z_{jt} \leq \psi z_{kt} \qquad (6.2.2)$$

$$\sum_{j=1}^{N} \lambda_j Y_j \geq Y_k ; \Sigma \lambda_j = 1 = \Sigma \mu_j$$

$$\lambda_j, \mu_j \geq 0$$

and

$$\text{Min } q'x + z$$

$$\text{s.t.} \quad \sum_j \lambda_j X_j \leq x ; \sum_j \mu_j z_{jt} \leq z ; \sum_j \lambda_j Y_j \geq Y_k \qquad (6.2.3)$$

$$\Sigma \lambda_j = 1 = \Sigma \mu_j ; (\lambda, \mu) \geq 0$$

Here capacity $z$ is a scalar variable, $(X_j, Y_j)$ are input output vectors for unit j and $q'$ is a row vector denoting unit costs (prices) for the variable inputs x. Denote by asterisks the optimal values of the decision variables. Then unit k is relatively inefficient in the use of capacity inputs if $\psi^* < 1$, whereas it is inefficient in the use of current inputs of $\theta^* < 1.0$. The optimal values $x^*$, $z^*$ of current and capacity input may also be compared with the actual levels $X_k$, $z_k$ used by unit k in order to locate efficient gap if any. In case market price data (p) are available for the output vector y and a two-period framework is assumed, then the optimal inputs and outputs can be determined from the LP model as follows:

$$\underset{y,x,z}{\text{Max}} \ \pi = p'_t y - q'_t x - w'_t z + (1+r)^{-1} w'_{t+1} \hat{z}$$

$$\text{s.t.} \qquad X\lambda \leq x; Y\lambda \geq y; \mu' z_t \leq z; \mu' \hat{z}_t \geq \hat{z} \qquad (6.2.4)$$

$$\lambda' e = 1 = \mu' e; \lambda \geq 0, \mu \geq 0$$

Here $z_t$ is a vector of durable inputs purchased at the beginning g of period t at prices $w_t$, $\hat{z}_t$ is a vector of depreciated durable inputs that will be available to the firm at the beginning of the subsequent period, $w_{t+1}$ is the vector of durable input prices that the firm anticipates will prevail during period t+1, and r is an appropriate discount rate exogenously given. Here the capacity-related inputs are the durable inputs and their unit costs are the input prices. With observed values $(X, Y, z_t, \hat{z}_t)$ of inputs and outputs the firm could now determine the optimal inputs and outputs $(x^*, y^*, z^*, \hat{z}^*)$. We note however some basic differences of this formulation from the traditional DEA models. First of all, the vector of spit prices $w_{t+1}$ is not observed at time t and hence the producer's anticipation of future price is needed. In this sense this model yields anticipated or expected efficiency. Since the anticipated prices are uncertain, the firm's attitude towards uncertainty must be modeled. This is the framework where the rational expectations (RE) hypothesis may be introduced. Secondly, the durable inputs are used here to approximate the stock of capacity inputs, but for certain stocks like natural resources and goods inventories there may be no natural market prices. Finally, the relevant discount rate r must be common to all the firms and also know. In the static DEA models these basic questions are not addressed at all.

In case of stochastic demand $d_t$ the DEA model (6.2.4) can be transformed as:

$$\text{Max E} \ \pi = E[p'_t \min(y, d_t) - q'_t x + (1+r)^{-1} \hat{z} - w'_t z]$$

$$\text{s.t.} \qquad \text{the same constraint as in (6.2.4)}$$

where E denotes expecation. On using the Lagrangian expression:

$$L = E\pi + \beta'(x - X\lambda) + \alpha'(Y\lambda - y) + \gamma'(z - \mu' z_t) + \delta'(\mu' \hat{z}_t - \hat{z})$$

$$+ \beta_0(1 - \lambda' e) + \gamma_0(1 - \mu' e)$$

We must have for the efficient unit:

$$p_t(1 - F(y^*)) - \alpha^* = 0$$

$$\gamma^* = w_t; \beta^* = q_t; \delta^* = (1+r)^{-1}$$

$$Y'\alpha^* = X'\beta^* + \beta_0^* e; \gamma^{*'} z_t = \delta^{*'} \hat{z}_t - \gamma_0^* e$$

This shows that the unit exhibits output inefficiency if $\sum\limits_{j=1}^{N} Y_j \lambda_j^* > y^*$, input efficiency if $\Sigma X_j \lambda_j^* < x^*$, capacity inefficiency if $\mu^{'*} z_t < z$ or $\mu^{*'} z_t > \hat{z}^*$. Clearly there are five sources of inefficiency in this framework: the input, output, capacity and inefficiency due to market demand uncertainty. The theory of organizational slack deals specifically with the demand and capacity oriented sources of inefficiency which may apparently inflate the marginal costs.

In case we have a time horizon it is simpler to introduce investment variables denoted by a vector $I_t$ and rewrite the long run profit function as:

$$\text{Max} \, E \left\{ \sum_{t=0}^{\infty} (1+r)[p_t' \min(y_t, d_t) - q_t' x_t - \rho_t' I_t - w_t' z_t] \right\}$$

$$\text{s.t.} \quad X_t \lambda_t \leq x_t; Y_t \lambda_t \geq y_t; I_t \leq (1+\delta_0) z_{t+1} - z_t$$

$$\lambda_t' e = 1; \lambda_t \geq 0$$

where investment is constrained by changes in capacity inputs with $\delta_0$ denoting fixed rates of depreciation. The theory of adjustment costs which relates current production to capital stock and investment in new capital along with the variable inputs is implicit in this formulation and its implications have been discussed by Artus and Muet (1990) in an empirical framework and by Sengupta (1995a) in the DEA framework.

For public sector enterprises however the market prices of output are generally unavailable and the profit maximization objective does not apply, since these are not for profit organizations. Hence in this case we may restrict ourselves to the cost frontier alone and use the theory of adjustment costs to develop ad model of capacity utilization. Consider the production function

$$y = f(v, x)$$

of a firm, which produces a single output $y$ by means of the vector $v$ of variable inputs and the vector $x$ of service flows from the quasi-fixed inputs (i.e., these inputs are fixed in the short run but variable in the long run). Since the production function may exhibit increasing returns to scale, the usual profit maximization principle may not yield determinate results. Hence we adopt the cost minimization model, where in the short runs the firm minimizes variable costs

$q'v$ in the short run subject to the production constraint $y \leq f(v,x)$ where $x$ is fixed. This yields the short run cost function $C_v = g(y,q,x)$. Denoting by $w$ the vector of rental prices for the quasi-fixed inputs, the total cost $C = C_v + C_x$ may be defined with $w'x = C_x$ as the fixed cost. Capacity output $\hat{y}$ is now defined by that level of output for which total cost $C$ above is minimized, i.e.,

$$\hat{y} = h(q, x, w) \tag{6.2.5}$$

with the associated cost function for capacity output

$$\hat{C} = G(q, w, \hat{y}) \tag{6.2.6}$$

Two implications of this concept of optimal capacity output must be noted. One is that the capacity output $\hat{y}$ may be viewed as a point of tangency between the short and the long run average total cost curves. Secondly, one can now define the rate of capacity utilization as $u = y/\hat{y}$ where $0 \leq u \leq 1$. Morrison and Berndt (1981) used this type of a dynamic cost function model with a single quasi-fixed input called capital to estimate the patterns of capacity utilization of U.S. manufacturing over the period 1958-77 by using a regression model. One can also use a DEA model to specify a cost frontier as follows:

$$\text{Min } s_k^{\perp} + \hat{s}_k$$

$$\text{s.t.} \quad a_0 + \sum_i a_i q_{ij} + \sum_i b_i x_{ij} + dy_j + s_j = C_j$$

$$\alpha_0 + \sum_i \alpha_i q_{ij} + \sum_i h_i w_{ij} + \delta \hat{y}_j + \hat{s}_j = \hat{C}_j \quad j=1,2,\ldots,N$$

where $C_j$ and $\hat{C}_j$ are observed short run and long run costs for unit $j$ and the observed data consist of input and output pices and the two outputs. If unit $k$ is efficient then we must have $s_k^*$ and $\hat{s}_k^*$ to be zero implying full capacity utilization.

If the short and long run cost components can be separately obtained as $C_{ij}$ and $\hat{C}_{ij}$, then these could be used more directly to characterize DEA efficiency as follows:

$$\text{Min } \varepsilon + \zeta$$

$$\text{s.t.} \quad \sum_{j=1}^{N} C_{ij} \lambda_j \leq \varepsilon C_{ik} ; \Sigma \lambda_j = 1, \lambda_j \geq 0$$

$$\sum_{j=1}^{N} \hat{C}_{ij}\mu_j \leq \zeta \, \hat{C}_{ik} \, ; \Sigma \mu_j = 1, \mu_j \geq 0 \quad j=1,2,\ldots,N$$

In this framework unit k is efficient in the short run if $\varepsilon^* = 1.0$, but not efficient in the long run if $\zeta^*$ is less than one. The fact that some inputs are fixed in the short run makes it clear that the rate of capacity utilization may influence short and long run costs differently. Sengupta (1995a, 1999) has discussed some stochastic extensions of this type of model in an intertemporal dynamic framework.

From an economic viewpoint the most important source of excess capacity is due to a fall in market demand, i.e., demand uncertainty and the existence of excess capacity tends to inflate the short and the long run costs of output. For public sector enterprises the competitive market pressure is very weak, hence the probability of incurring dead weight losses and hence inefficiency due to organizational slack is much higher. The extensions of DEA models proposed here may thus provide a very useful interface with the economic theory of efficiency.

## 6.3 Cost Efficiency and FMS

Recently DEA models have been empirically applied to flexible manufacturing systems (FMS). Thus Shang and Sueyoshi (1995) have applied an interlinked set of DEA models to select the most appropriate FMS for a manufacturing organization. They prpose interlinking three individual modules: an analytic hierarchy process, a simulation module and an accounting procedure. In this section we propose a cost-based allocative efficiency model for selecting the most appropriate FMS setup. The objective of FMS are to improve both flexibility and productivity. Modelling for FMS geneally involves cost comparisons at various levels, e.g., at the plant, shop floor, division or system management. Recently, Sengupta (1995a) discussed three major types of FMS models which are directly relevant for DEA models. The first deals with productivity based models, where various types of flexibility, e.g., process flexibility, operation flexibility, expansion flexibility are introduced as components of the production and cost function in order to maximize its contribution to profits or output. The second type of model introduces the economies of scope argument in output expansion for technology-intensive plants or DMUs. Whereas the economies of scale refer to cost advantages due to increase in size, volume or scale, the conomies of scope refer to cost savings due to increasing complexity due to the varieties of products produced by the plants. The third type of model looks into the learning curve effects on cost reduction in a plant due to the accumulation of skills and experience of the workers in the plant. Recently, Fine and Freund (1986) have developed a model where the managers have to choose between two types of capacity (investment) for several products; one is a product flexible capacity the other is non-flexible. Since flexible capacity proves to be more cost-effective in the long run, a long run extension of the DEA model is appropriate here. Thus, any DMU which proves to be inefficient in the short run, due to grater use of more expensive flexible capacity, may indeed be more efficient in the long run due to its long run cost advantages. Several such cases have been discussed by Sengupta (1995a).

Another useful cost efficiency model for automated flexible production environments, which has recently been applied by Alberts et al. (1989), is most suitable for extended DEA analysis, since it involves the basic cost elements of an FMS system for which data are generally available in the manufacturing plants. Cost allocation, queueing network and the servicing discipline are the most used techniques in modelling the allocation process of the FMS resources to the parts processed. However, conventional costing techniques used in traditional discrete productions do not apply to a FMS environment. Three special features specific to this environment show the need for new procedures for manufacturing cost analysis, e.g. (a) *automation* which substantially modifies the role of labor costs; (b) *flexibility*, which emphasizes the need to distribute the cost of the use of FMS resources among all the parts that are concurrently processed; and (c) *integration* among the system resources, which leads to a higher influence on the operating strategy of the system performances. Since capital costs assume more importance in the total manufacturing costs, the utilization rate of the system resources plays a critical role in determining economies. Note that the conventional DEA models ignore this entire set-up of flexible technology including its structure of varying utilization rates. To illustrate the use of an extended DEA model in this set-up, we consider the cost efficiency model developed by Alberts et al (1989) as follows.

Let K be the capital and operating cost per time unit of the entire installation. If J part type is simultaneously processed and the estimated throughput of part type j (j=1,2,...,J) is $x_j$, then the unit manufacturing cost $c_{nj}$ of part type j must satisfy the condition $K = \sum_{j=1}^{J} c_{nj} x_j$. Now the technological operations involving machining, fixturing, washing, mixing, etc. involve dedicated resources (i.e. workstations) over periods of time which give rise to technological costs. Assume there are M dedicated resources when part type j requires technological operations on workstations i (i=1,2,...,I) with service time $t_{ij}$. If $c_i$ denotes the quota of capital and operating cost K imputable to workstation i, then the cost ($c_{ij}$) of technological operations is given by

$$c_{ij} = (c_i t_{ij})/u_i \qquad\qquad (6.3.1)$$

where $u_i$ ($0 < u_i < 1$) is the rate of utilization of workstation i (i=1,2,...,I) during the residence of part j (j=1,2,...,J) within the FMS.

Besides technological operations the FMS also contains logistical operations involving transfer, dispatching, waiting, etc, which engage common shared resources, e.g., hardware and software of the control system, supervision personnel etc. Let $c_S$ be the quota of total cost K and $c_{S_j}$ the cost of the logistic operations of part type j on the common shared resources, then

$$c_{S_j} = (c_S / n) \sum_{i=1}^{I} t_{ij} \qquad\qquad (6.3.2)$$

where $T_j = \sum_i t_{ij}$ is the residence time of part type j within the FMS and $n = \sum_{j=1}^{J} n_j$ is the total number of parts simultaneously present within the FMS and $n_j$ is the number of pallets of part type j. On combining the two types of costs specified above, one obtains the unit total manufacturing costs ($c_j$) of part type j as follows:

$$c_j = \sum_{i=1}^{I} \left( c_i t_{ij} / \sum_{r=1}^{J} t_{ir} x_r \right) + (c_S / n) \sum_{i=1}^{I} t_{ij}$$

$$= \sum_{i=1}^{I} (c_i / t_{ij} / u_i) + b_S T_j \qquad\qquad (6.3.3)$$

wehre $u_i = \sum_{j=1}^{J} t_{ij} x_j$ is the utilization rate of the dedicated resource used on workstation i, $T_j = n_j / x_j = \sum_j t_{ij}$ and $b_S = c_S/n$.

Clearly one could define in this framework three measures of cost efficiency. One is the utilization rate $u_i$ $(0 < u_i \le 1)$. If $u_i = 1.0$ for each i, then the cost of part j becomes

$$c_j^* = \sum_i c_i t_{ij} + b \sum_i t_{ij} = \sum_i (c_i + b_S) t_{ij} \qquad\qquad (6.3.4a)$$

where $c_j^* < c_j$ for $0 < u_i < 1$ for at least one i and $c_j$ is the observed total manufacturing cost per part type j. A second measure of efficiency is defined in terms of the DEA model:

$$\left.\begin{array}{c} \text{Min } \varepsilon_k = c_k - \sum_{i=1}^{I} (c_i + b_s)t_{ik} \\[2mm] \text{s.t.} \sum_{i=1}^{I} (c_i + b_S)t_{ij} \le c_j; \\[2mm] j = 1, 2, ..., J \\[2mm] t_{ij} \ge 0; \\[2mm] j = 1, 2, ..., J \end{array}\right\} \qquad (6.3.4.b)$$

Here, part k is produced efficiently, if the optimal solution $(t_{ik}^*)$ of the above model (6.3.4) satisfies the equality condition

$$c_k^* = \sum_i (c_i + b_S)t_{ik}^* = c_k$$

and all the slack variables are zero. In this case $c_k^* = c_k$. In the case where part k is produced inefficiently, then

$$c_k^* = \sum_i (c_i + b_S)t_{ik}^* < c_k \qquad (6.3.5)$$

must hold.

Finally, the overall system efficiency can be measured by a single index representing the cost efficiency index (CEI) as follows:

$$\text{CEI} = \sum_{j=1}^{J} c_j^* x_j / (\sum_{j=1}^{J} c_{nj} x_j) = \sum_{j=1}^{J} c_j^* x_j / K \qquad (6.3.6)$$

where $c_{nj} = \sum_{i=1}^{I} c_{ij} + c_{Sj}$ with $c_{ij}$ and $c_{Sj}$ defined by equations (6.3.2) and (6.3.3) respectively. Denote the minimum value of CEI in (6.3.6) by $\phi^* = \phi_h^*$ for plant h. A comparison of different plants $\phi_1^*, \phi_2^*, ..., \phi_H^*$ can then be performed by arranging these efficiency indices in an ascending order $\phi_{(1)}^* \le \phi_{(2)}^* \le ... \le \phi_{(H)}^*$. In this case the most efficient plant is characterized by $\phi_{(1)}^*$.

Two important comments may be made about the cost efficiency analysis above. One is that it integrates the two aspects of efficiency analysis: the technological and the economic. Secondly, the residence time $t_{ij}$ which are observed data in this framework can be replaced by the estimates of mean residence times $r_{ij}$ by assuming that the service times are exponentially distributed. Alberts *et al* (1989) have used the following estimating equation for ($r_{ij}$):

$$r_{ij} = t_{ij}\left(1 + \frac{n_j - 1}{n_j}(q_{ij} - 0.5u_{ij})\right)$$

$$+ \sum_{\substack{g=1 \\ g \neq j}}^{J} t_{ij}(q_{ig} - 0.5u_{ig})$$

where $q_{ij}$ is the mean number of part j at workstation i and $u_{ij}$ is the mean utilization rate of station i by part type j. This formulation allows a relaxation of the assumption of deterministic service times.

## 6.4 Sequential Models of Production Planning

The DEA models of production efficiency may be developed into separable stages and then interlinked sequentially depending on the objectives of the model builder or the policy maker. Thus one may consider a set of firms, then a set of industries and finally the whole economy. Altenatively, one may use panel data to determine first the set of sample points which are cost efficient and then use the parmaeters of the cost frontier in selecting optimal levels of inputs for the future planning period. We consider these two cases and show its connection with the general equilibrium model of DEA efficiency.

For analyzing industry efficiency let us consider for example a specific industry such as coal, which has been analyzed for both productoin planning (see e.g., Henderson (1956)) and DEA efficiency analysis (see e g., Byrnes, Fare and Grosskopf (1984)).

Consider a three-stage model for the coal industry, incorporating both demand, costs and a heterogeneous set of coal mines. At the first stage one specifies a macro model for the whole conomy, where the output is produced in state or district i and shipped to destination or consuming centers j. Let $x_{ij}$ dente the shipment quantity where $X_i = \sum_{j=1}^{n} x_{ij}$ is total output at i and $Y_j = \sum_{i=1}^{m} x_{ij}$ is total utilization at centre j, where $i \in I_m, j \in I_n$ and $I_p = \{1,2,\ldots,p\}$ is he index set. Denoting extraction or production costs by $c(X_i)$ and transportation costs by $t(x_{ij})$ a cost minimizing model can then be specified as follows:

$$
\left.
\begin{array}{l}
\min C = \sum_{i=1}^{m} c(X_i) + \sum_{i=1}^{m} \sum_{j=1}^{n} t(x_{ij}) \\[2mm]
\text{subject to} \\[2mm]
\quad Y_j \geq D_j (\text{demand}) \\[1mm]
\quad X_i \leq K_i (\text{capacity}) \\[1mm]
\quad x_{ij} \geq 0; i \in I_m, j \in I_n
\end{array}
\right\} \qquad (6.4.1)
$$

Assuming linear extraction and transportation costs, Henderson (1956) applied such a model for the U.S. economy by grouping the states into 14 continuous districts and by averaging the quantities of demand and capacity for the three years 1947, 1949, and 1951 with two types of deposits identified as surface (or open cast) and underground deposits. The overall percentages of these two types of deposits in 1951 were, for example, 73.7 and 2.3, whereas the total demand for coal was 1.3m, 0.98m and 1.08m in units of 10 billion btu respectively for 1947, 1949, and 1951. Coal is extracted in 11 out of 14 districts and hence m = 22 with two types of deposits and n = 14 since coal is consumed in all 14 districts. The estimate of capacity ($K_i$) used here is defiend as the maximum amount of coal that can be extracted in a given year with the labor and equiment available during that year.

Two types of modifications are in order for recent periods due to many technical changes in the coal industry. One is the relative decline in unit extraction costs due to modernization and technical change. Also, both demand and capacity estimates have frequently undergone revisions due to marked cometition from other industris and other regulatory reforms related to environmental safeguards etc. Secondly, output prices have also changed due to competition from the other industries in the power sector. To capture this aspect of the imperfectly cometitive market one may postulate a profit function rather than the cost function as the objective, i.e.

$$
\max \pi = \sum_{j=1}^{n} p_j Y_j - \sum_{i=1}^{m} (c_i x_{ij} - (\gamma_i / 2) x_{ij}^2) - \sum_{i=1}^{m} \sum_{j=1}^{n} t_{ij} x_{ij} \qquad (6.4.2)
$$

where the extraction cost parameters are $c_i$, $\gamma_i$ and $t_{ij}$ denotes the transport costs per unit of shipment of coal. With a linear demand function

$$
p_j = a_j - b_j Y_j
$$

the final form of the objective function reduces to

$$\pi = \sum_{j=1}^{n} (a_j Y_j - b_j^2 Y_j^2) - \sum_{i=1}^{m} \left[ c_i x_{ij} - (\gamma_i / 2) x_{ij}^2 + \sum_{j=1}^{n} t_{ij} x_{ij} \right] \qquad (6.4.3)$$

Let $x_{ij}^*$ be the optimal solution of the quadratic program (6.4.3) with a positive value, then it follows by the Kuhn-Tucker theorem that

$$MR_j + \lambda_j^* - \mu_j^* = MC_j \qquad (6.4.4)$$

where

$$MR_j = \text{marginal revnue} = a_j - 2b_j Y_j^*$$
$$MC_j = \text{marginal cost} = c_i - \gamma_i x_{ij}^* + t_{ij}$$
$$\lambda_j^* = \text{imputed price of demand constraint}$$
$$\mu_i^* = \text{imputed royalty of capacity utilization.}$$

In the linear case of cost minimization, the positive optimal solution $x_{ij}^*$ satisfies the equality

$$\lambda_j^* - \mu_j^* = MC_j = c_i + t_{ij} \qquad (6.4.5)$$

However, if there is less than full capacity utilization for any deposit i, then $\mu_i^*$ is zero and one obtains the equality of imputed price with marginal cost of extraction and transport

$$\lambda_j^* = c_i + t_{ij} \qquad (6.4.6)$$

Thus, it is clear that this type of LP or QP (quadratic programming) model yields three types of efficiency measures:

(1)     the divergence of actual quantities $x_{ij}$ from the optmal quantities $x_{ij}^*$;
(2)     ranking of deposites or mining districts in terms of the Lagrange multiplier $\mu_i^*$, which represents the imputed royalty due to capacity utilization; and
(3)     the divergence of actual profit ($\pi$) from the optimal profit ($\pi^*$) or actual total cost (C) from the optimal cost ($C^*$).

The second stage decision problem is to choose an efficient set of mines for each efficient district i supplying the optimal quantity $X_i^*$. Denote the outputs of N mines by $X_{ir}$, $r \in I_N$ and the various inputs by $I_{kr}$, $k \in I_{S_i}$ where there are $S_i$ inputs for each i. For testing the relative efficiency of each mine belonging to the ith district, we set up a DEA model for each reference mine and test if it is efficient. For example if the hth mine is the reference unit, one sets up the following model

$$\text{Min } g_h = \sum_{k=1}^{S_i} \beta_k I_{kh}$$

subject to

$$\sum_{k=1}^{S_i} \beta_k I_{kr} - \alpha X_{ir} \geq 0 \qquad (6.4.7)$$
$$r \in I_N ; \alpha X_{ih} = 1$$
$$\beta_k \geq 0, \alpha \geq 0$$

By dropping the subscript i for convenience and using a = $1/X_{ik}$ on could rewrite it as a simpler LP model

$$\text{Min } g_h = \sum_k \beta_k I_{kh}$$

subject to

$$\sum_k \beta_k I_{kr} \geq \hat{X}_r, \hat{X}_r = X_{ir} / X_{ik} \qquad (6.4.8)$$
$$\beta_k \geq 0; r \in I_N$$
$$k \in I_{S_i}$$

where $\hat{X}_r$ is the relative output level of unit r expressed in terms of the reference unit's output level. By varying h over the index set $I_{S_i}$, one obtains $S_i$ LP models for each i. The otpimal solution of these LP models determines two subsets of mines, one containing the efficient units and the other the inefficient ones.

The dual of the LP model (6.4.7) used to test the relative efficiency of the mining unit h can easily be computed as follows:

$$\text{Max } \theta$$
subject to

$$\sum_{r=1}^{N} \lambda_r I_{kr} \leq I_{kh}$$

$$\sum_{r=1}^{N} \lambda_r I_{ir} \geq \theta X_{ih} \qquad (6.4.9)$$

$$\lambda_r \geq 0 \; ; \; \theta \text{ free in sign}$$

$$k \in I_{S_i}, r \in I_N \text{ and } i \in I_m$$

Let $(\lambda_r^*, \theta^*)$ be the optimal solution of this LP model and denote the left-hand side of the first two constraints as $I_{kh}^* = \sum_r \lambda_r^* I_{kr}$ and $X_{ih}^* = \sum_{r=1}^{N} \lambda_r^* X_{ir}$. These are potential minimum inputs and potential maximum outputs for the reference unit h. Clearly, if each $X_{ir}$ is positive, then the optimal value of $\theta$ is positive. Also, if it holds for the mining unit h that

either $I_{kh}^* < I_{kh}$, for at least one k

or $X_{ih}^* > X_{ih}$ for least one i $\qquad (6.4.10)$

or both

then the unit h cannot be technically efficient, since there exist potneital inputs lower, or potential outputs higher. Thus, the reference unit h is efficient, if and only if the following two conditions are satisfied:

$\theta^* = 1.0$.

all slacks in LP (6.4.9) are zero at the optimum.

The non zero slacks and the value of $\theta^* < 1$ identify the source and amounts of any inefficiencies that may be present. Several variants of the DEA formulations (6.4.7) and (6.4.8) above have been discussed by Charnes et al (1994). The case of multiple outputs identified by different grades of coal can be easily incorporated in this framework.

It is clear that the value of $\theta^*$ in an ascending order may be used to characterize two sets of mines, one efficient and the other inefficient. However, two problems still remain. One is that all the efficient units combined may produce a total output less than the optimal quantity $X_i^* = \sum_{j=1}^{n} x_{ij}^*$ determined in the first stage. Secondly, the management policy must be designed so as to improve the inefficient units by either reduction of inputs or by merger of selected inefficient inputs. To handle these problems we introduce the third stage of an allocation-type LP model as follows:

$$\max \sum_{r=1}^{N} X_{ir}$$

subject to

$$\sum_{r=1}^{N} a_{kir} X_{ir} \leq A_{ki} \qquad\qquad (6.4.11)$$

$$\sum_{r=1}^{N} X_{ir} \geq X_i^*$$

$$I \in I_m; K \in I_{S_i}; r \in I_N$$

Here $X_{ir}$ is the new deicsion variable, $A_{ki}$ is the aggregate input of type k for each unit i and akir is the input-output coefficient for input type k and output of the rth unit at district i. We have m LP models of type (6.4.11) which ensures that the efficient outputs satisfy the overall demand constriants, given the technology matrix $(a_{kir})$ for each i and the aggregate input vector. Let $X_i^{**} = \sum_{r=1}^{N} X_{ir}^{**}$ be the optimal solution of the LP model (6.4.11). This optimal solution $X_i^{**}$ has several flexible features that are suitable for managerial planning and control. First of all, it satisfies the demand constraints specified at the first stage. Hence, if the demand for coal changes, it may necessitate a change in the level of $X_i^*$ and hence a change in $X_i^{**}$. Thus, the effect of falling demand on capacity utilization may be easily tracked down in this three-stage model. Secondly, if some inputs like capital equiment is not fully utilized in the first set of constraints (6.4.11), it will show inefficiency in capital utilization. Thus, it provides for input efficiency profiling, which has been suggested as a method to supplement the traditional DEA model, e.g., Tofallis (1997) has used this input efficiency profiling technique for efficiency evaluation of international airlines. However, the method proposed in (6.4.11) is more geneal, since it can handle all the inputs simultaneously. Finally, if the management proposes a reward or incentive system, its impact on the optimal output can be directly assessed here. The dual to this model (6.4.11) has the usual interpretation in terms of shadow prices of the inputs. Furthermore, if new technology is introduced through modernization, it can be analyzed through different vintages of output $(X_{ir}^v)$ and its coefficients $(a_{kir}^v)$, where the superscript v denots a vintage.

Thus, the three-stage version of the industry efficiency model supplements the DEA efficiency analysis through two additional LP models that are interlinked.

Next consider a two-stage allocative DEA model which can be applied in production planning problmes. For this purpose we consider the aggregate production planning model known as the HMMS model originally formulated by Holt et al (1960) that has been extensively applied in management science literature. Following this model we define the following cost components:

$$C_{1t} \geq C_{1t}^* = c_1 x_{1t} \text{ ( regular payroll)}$$

$$C_{2t} \geq C_{2t}^* = c_2(x_{1t} - x_{1,t-1})^2 \text{ (hiring and layoff)}$$

$$C_{3t} \geq C_{3t}^* = c_3(x_{2t} - c_4 x_{1t})^2 + c_t x_{2t} - c_6 x_{1t} \text{ (overtime and idle time)}$$

$$C_{4t} \geq C_{4t}^* = C_7^*(x_{3t} - c_8 - c_9 d_t)^2 \text{ (inventory and shortages).}$$

where $C_{it}$ and $C_{it}^*$ are the observed costs and minimal costs for different components and $x_1$ = volume of workforce, $x_2$ = output, $x_3$ = inventory and $d_t$ is the estimated sales. Our samples are $\tau$ periods ($t = 1,2,...,\tau$) for which the data set of $C_{it}$ and $x_{jt}$ are observed. To test which of the samle points are technically efficient we solve the nonlinear DEA model

$$\min_c \sum_{i=1}^{4} (C_{ik} - C_{ik}^*) = \max \sum_{i=1}^{4} C_{ik}^*$$
subject to
$$C_{it} \geq C_{it}^*, \text{ I=1,2,3,4; t=1,2,...,}\tau \qquad (6.4.12)$$

for observation at kth time point ($k=1,2,...,\tau$). By varying k in the objective function over the sample set ($k=1,2,...,\tau$), one may test the DEA efficiency for each point. Let E be the set of DEA-efficient observations out of the $\tau$ sample points. Let $c^* = (c_1^*,...,c_9^*)$ be the parameters estimated for any one of the DEA-efficient observations. Given $c^*$, one may then set up the dynamic decision model

$$\min_x C_T = \sum_{\tau=1}^{T} \sum_{i=1}^{4} C_{i\tau}^*(x_1, x_2, x_3) \qquad (6.4.13)$$

in the second stage for determining the optimal levels of $x_{1t}$, $x_{2t}$ and $x_{3t}$ over the planning period $t = 1,2,...,T$. Note that this two stage model is basically the same as beore with two differences. One is that it is a nonlinear cost frontier, so that the method of nonlinear programming has to be applied here to obtain the DEA parameters $c^* = (c_1^*,...,c_9^*)$. Secondly, the demand or sales variable $d_t$ has to be estimated or predicted when the second stage model (6.4.13) is applied, since there is an inventory constraint in the HMMS model as

$$x_{3t} = x_{3,t-1} + x_{2t} - d_t \geq 0.$$

Otherwise, demand or sales has to be modelled as a stochastic process and the objective function in (6.4.13) has to be modified as the minimization of an expected cost function: min $EC_T$.

Seveal flexible features of this type of formulation are to be notd. One is that it allows sequential updating of the optimal decisoin rule yielded by the second stage model (6.4.13) as the time horizon T is shifted forward over time. Secondly, the selection of the parameter vector c* at the first stage may be determined by the mode or the median, as is done in DEA literature. Finally, the optimal linear decision rules (LDR) of the model (6.4.13) can be easily computed in the form

$$x_{1t} = \sum_{i=0}^{t-1} \alpha_i d_{t+1} + \sum_{i=1}^{t-1} \beta_i x_{1,t-i} + \sum_{i=1}^{t-1} \gamma_i x_{3,t-i} + \alpha_0$$

$$x_{2t} = \sum_{i=0}^{t-1} a_i d_{t+1} + \sum_{i=1}^{t-1} b_i x_{1,t-i} + \sum_{i=1}^{t-1} g_i x_{3,t-i} + a_0 \qquad (6.4.14)$$

$$x_{3t} = x_{3,t-1} + x_{2t} - d_t,$$

where parameters $\theta = (a_i, b_i, g_i, a_0, \alpha_i, \beta_i, \gamma_i, \alpha_0)$ are functoins of the original paameters $c^* = (c_1^*, ..., c_9^*)$ estimated by the DEA efficient observations. Since the optimal values of the decision vector $x_t = (x_{1t}, x_{2t}, x_{3t})$ depend on future demand $d_{t+1}$, it is clear that future changes in demand would affect the optimal values of $x_i$ as conditional forecasts $E(x_t|I_t)$ given the information set $I_t$ available up to the time point t are used in the updating rules. Thus, demand fluctuations which may induce variations in capacity and inventory utilization rates may be directly incorporated in this formulation. An empirical application of this model is discussed in Chapter 5.

Finally, we consider a general equilibrium model for the whole economy, where Pareto efficiency conditions hold for production, distribution and consumption simultaneously.

We asusme H consumers who are final users of commodities denoted by an r-vector $z_h$ (h=1,2,...,H) and owners of an initial endowment of commodities $e_h$. The set of all feasible consumption bundles is denoted by $Z_h$. There are J producers where each producer j (j=1,2,...,J) produces a production vector $y_j$ with r commodities, where we adopt the convention that positive components are outputs and negative components inputs. The set of feasible productoins is denoted by $Y_j$. We introduce two more concepts to characterize a competitive market equilibrium:

(1) prices represented by a nonnegative r-vector p

(2) coefficients $\theta_{hj} \geq 0$ with $\sum\limits_{h} \theta_{hj} = 1$ denoting the share of profits of producer j

      occurring to consumer h

Now consider a feasible allocation $(z_h, y_j)$ over appropriate feasible sets where h=1,2,...,H and j=1,2,...,J. This allocation is Pareto efficient if it satisfies each of the three conditions of efficiency as follows:

(1) efficiency in productoin: if it is not possible to choose alternative production plans $\hat{y}_j$ yielding more output from the same or less input or yielding the same output from less input,

(2) efficiency in consumption: if it is impossible to redistribute the given totals of total goods produced and of goods used as inputs differently among consumers so as to make some of them better off and none worse off and,

(3) efficiency in coordination of production and consumption: if it is not possible to make some consumers or producers better off and none worse off.

Competitive equilibrium is specified by a price vector $\bar{p}$, if it satisfies the following three conditions:

(a) equality of demand and supply for nonfree goods:

$$\Sigma \bar{z}_h - \Sigma \bar{y}_j - \Sigma e_h \leq 0$$
$$\bar{p}'(\Sigma \bar{z}_h - \Sigma \bar{y}_j - \Sigma e_h) = 0 \qquad\qquad (6.4.15)$$

(b) utility maximization by consumers:

$\bar{z}_h$ maximizes $U_h(z_h)$

s.t. $\qquad \bar{p}'\bar{z}_h \leq \bar{p}'e_h + \sum\limits_{i} \theta_{hi}\bar{p}'\bar{y}_i \qquad\qquad (6.4.16)$

$\qquad\qquad z_h \in Z_h$

and

(c) profit maximization by firms:

$\bar{y}_j$ maximizes $\bar{p}'y_j$    s.t.   $y_j \in Y_j$ $\qquad\qquad\qquad\qquad (6.4.17)$

Under standard regularity conditions on the sets $Z_h$ and $Y_j$ (e.g., nonincreasing returns to scale, free disposal, positive initial endowment) and quasi-concavity of the utility function $U_h$ it can be shown that there exists an equilibirum price vector $\bar{p}$. Hence every competitive equilibrium implies a Pareto optimum or efficient allocation.

It is clear that the production efficiency aspect of Pareto optimum may be analyzed by a DEA model as

Max $p'y$

s.t. $\quad \sum_k y_k \lambda_k \geq y; \Sigma \lambda_k = 1, \lambda_k \geq 0$ $\qquad$ (6.4.18)

If input vectors $x_k$ are separately introduced with prices q, then this model becomes

Max $\pi = p'y - q'x$

s.t. $\quad \sum_k y_k \lambda_k \geq y; \sum_k x_k \lambda_k \leq x; \Sigma \lambda_k = 1, \lambda_k \geq 0$ $\qquad$ (6.4.19)

For consumption efficiency one may set up the following model in linear utility functions:

Max $U_h = w'z$

s.t. $\quad \sum_k z_k \lambda_k \geq z; p'(z - e_h) \leq \sum_j \theta_{hj} p' y_j$ $\qquad$ (6.4.20)

$\qquad \Sigma \lambda_k = 1, \lambda_k \geq 0$

Note that the two models (6.4.18) and (6.4.20) are interlinked in two ways: one is that both consumers and producers are price takers and secondly the market equilibrium satisfies the Walrasian condition of excess demand (supply) i.e. the equilibrium price satisfies the two conditions:

Nonpositive excess demand:      $\Sigma z_h - \Sigma y_j - \Sigma e_h \leq 0$

Zero price for excess supply:      $p'(\Sigma z_h - \Sigma y_j - \Sigma e_h) = 0$

Thus one may conclude that the allocative DEA model (6.4.18) must use appropriate prices p according to competitive equilibrium in order to achieve the result that productoin efficiency also implies efficiency in consumption and the trading process involving the efficient coordination of consumption and production. Also, one has to note that if the competitive market prices change due to the overall macrodynamic conditions of the economy, production efficiency measures of the DEA models have to change accordingly. Thus market distortions or imperfectly competitive market conditions would necessitate important modifications of the DEA efficiency rule.

## 6.5 Efficiency under Risk Avesion

In DEA models of allocative efficiency the input price fluctuations affect the efficiency frontier through the risk attitudes of the DMUs. Thus the optimal input vector x is determined by the DMU by maximizing the expected von Neumann type utility function as

$$\text{Max} - EU(q'x)$$

$$\text{s.t.} \quad \sum_{j=1}^{N} x_{ij}\lambda_j \leq x; \sum_{j=1}^{N} y_{rj}\lambda_j \geq y_{rk}; i \in I_m; r \in I_s \qquad (6.5.1)$$

$$\Sigma\lambda_j = 1, \lambda_j \geq 0, j \in I_N$$

For the class of utility functions with a constant rate of absolute risk aversion the objective function here may be replaced by a risk adjusted disutility or cost function:

$$\text{MinC} = \overline{q}'x + (w/z)x'Vx; w \geq 0 \qquad (6.5.2)$$

where the random input pice vector q is assumed to be distributed with mean $\overline{q}$ and variance-covariance matrix V and w represents the degree of risk aversion. If $x^*$ is the optimal input vector solved from this quadratic program and $DMU_k$ is efficient, then we would have the following efficiency condition satisfied:

$$\beta^* = wVx^* + \overline{q}$$

$$\alpha^{*'} Y_k = \beta^{*'} X_k + \beta_0^*$$

where the Lagrangian is $L = -\overline{q}'x - (w/2)x'Vx + \beta'(x - \sum_j X_j\lambda_j)$

$+\alpha'(\sum_j Y_j\lambda_j - Y_k) + \beta_0(1 - \Sigma\lambda_j)$ and $X_j$, $Y_j$ are the input and output vectors of $DMU_j$. With no risk aversion w=0 and hence the mean input costs would equal $\beta^{*'} x^*$; otherwise higher price variance would increase the optimal input costs $\beta^{*'} x^* = \overline{q}'x^* + wx^{*'} Vx^*$. If the empirical estimates of $\overline{q}$, V can be made from observed data, then this type of model can be easily applied to characterize the risk averse cost frontier, provided all DMUs have the same degree of risk aversion measured by w.

A typical area where this type of risk averse efficiency frontier can be empirically applied is the capital asset market, where market efficiency is usually characterized by the capital asset pricing model (CAPM). For example DMUs here would be the various mutual funds, where the inputs are loads, turnover ratio, expense ratio, and the standard deviation of returns for riskiness of the portfolio. The outputs are mean returns for different funds. A typical input-oriented DEA model may then be specified by

$$\text{Min } \theta$$

$$\text{s.t.} \quad \sum_{j=1}^{N} x_{ij}\lambda_j \leq \theta x_{ik}; \sum_j y_j\lambda_j \geq y_k; \Sigma\lambda_j = 1 \qquad (6.5.3)$$

$$\lambda_j \geq 0; i \in I_m, j \in I_N$$

If most of the mutual funds (for example 75% or above) are efficient in terms of the this DEA, then the mean variance efficiency condition of the CAPM model is broadly satisfied. An empirical application of this type of formulation has been discussed in Chapter 5.

An alternative way to view the mean variance model of comparing the performance of mutual funds is to adopt the expected utility function approach in (6.5.1). On taking the risk averse form (6.5.2) one could specify an allocative DEA model as follows:

$$\text{Max } \bar{y} - (w/2)\sigma^2$$

$$\text{s.t.} \quad \sum_j x_{ij}\lambda_j \leq x_{ik}; \sum_j \sigma_j^2\lambda_j \leq \sigma^2 \qquad (6.5.4)$$

$$\sum_j \bar{y}_j\lambda_j \geq \bar{y}; \Sigma\lambda_j = 1, \lambda_j \geq 0$$

where the input riskiness of fund j has been separately written as $\sigma_j^2$. The efficient fund k would then satisfy the following conditions

$$\alpha^* \bar{y}_k = \beta^{*\prime} X_k + \beta_0^* + \sigma_k^2$$

$$\alpha^* = 1; b^* = w/2 \text{ if } \bar{y} \text{ and } \sigma^2 > 0$$

where the Lagrangian function is $L = \bar{y} - (w/2)\sigma^2$

$$+\beta'(x_k - \Sigma X_j\lambda_j) + b(\sigma^2 - \Sigma\sigma_j^2\lambda_j)$$

$$+\alpha(\Sigma\bar{y}_j\lambda_j - \bar{y}) + \beta_0(1 - \Sigma\lambda_j) \text{ and } \bar{y}, \sigma^2 \text{ are the mean and variance to be}$$

optimally determined. Let $\bar{y}^*, \sigma^{2*}$ be the optimal values of the efficient mutual fund. The closeness of all other funds' means and variances $(\bar{y}_j, \sigma_j^*; j = 1, 2, ..., N)$ would then indicate their efficiency rankings.

Two other aspects of efficiency evaluation are of some importance in stochastic frameworks. One is the dynamic extension of the portfolio efficiency frontier discussed above. The second is the concept of structural efficiency, which Farrell discussed in relation to inter-industry comparisons. These two are now discussed.

## A. Dynamic Portfolio efficiency Frontier

The specification of a dynamically optimal portfolio policy in a multiperiod framework offers some interesting generalizations to the one-period model of portfolio management, e.g., it may show that a suitable multiperiod policy may imply a myopic policy in suitable situations.

We consider here a utility based asset pricing model in an intertemporal horizon which can be easily related to the mean variance approach. We assume the investor j to be maximizing the expected utility of his final wealth $E\{U(W_{jt})\}$ where the wealth is defined by a discrete-time control system

$$W_{j,t+1} = s_{jt}W_{jt} + \sum_{i=1}^{n} r_{ijt}u_{ijt}; t = 0,1,...,T-1 \qquad (6.5.5)$$

where $s_{jt}$ and $r_{ijt}$ are the rates of return of the riskless asset and the risky asset and $u_{ijt}$ is the proportion invested in risky asset i at the beginning of period t. We assume for simplicity that $s_{jt} = s$ to be nonrandom and $r_{ijt} = (r_{it}) = r_t$ to be random with a mean vector $m_t = (m_{it})$ and variance covariance matrix $V_t$. Thus we can write the system dynamics above as

$$W_{j,t+1} = sW_{jt} + r_t'u_t + \varepsilon_t \qquad (6.5.6)$$

where $u_t = (u_{it})$ is the control vector and the error term $\varepsilon_t$ is assumed to be identically and independently distributed with a zero mean and fixed variance. As for the utility function we assume a quadratic form $U(W) = W—kW^2$ in conformity with the Markowitz-Tobin mean variance model. The DEA model thus becomes:

$$\text{Min } E(W_T^2)$$

$$\text{s.t.} \quad E(W_T) \geq c_T$$

$$\sum_{j=1}^{N} W_{j,t+1}\lambda_{j,t+1} \geq W_{t+1}, t = 0,1,...,T-1$$

$$\sum_j \lambda_{j,t+1} = 1, \lambda_{jt} \geq 0; \sum_i u_{it} = 1, u_{it} \geq 0$$

This formulation compares N investors and computes the optimal levels of $W_t^*, u_t^*$ and $\lambda_{jt}^*$ for t=0,1,...,T-1. This computation involves control theory algorithms and dynamic programming. Since

$$EW_T^2 = E[E(W_T^2 \mid W_0, W_1,..., W_{T-1}; u_0, u_1,..., u_{T-1}]$$

where the outer expectation is taken with respect to the random variables $W_0, W_1,..., W_{T-1}$ which depend on vector $r_t$. Clearly $EW_T^2$ is minimized by minimizing the inner conditional expectation with respect to $u_{T-1}$ for every possible collection of $(W_0, W_1,..., W_{T-1}; u_0,..., u_{T-1})$. Thus in the special case where the investor k is efficient with equality holding in the constraints above we obtain the optimal linear decisoin rule (LDR) as

$$u_{T-1}^* = R_t^{-1} m_t (c_T - s W_{t-1}) / \alpha_t$$
$$R_t = E(r_t r_t'), \alpha_t = m_t' R^1 m_t$$

and the associated minimal variance of $W_T$ as

$$\sigma_T^{2*} (c_T - s W_{t-1})^2 (1 - \alpha_t) / \alpha_t$$

where $W_{T-1} = h_{T-1}(W_0)$ can be specified as a suitable nonlinear function of the initial value $W_0$ by the dynamic programming algorithm. In some cases when the yield vector $r_t$ has identical probability distributions in all periods and the utility function satisfies the conditon of linearity of the risk tolerance function, then the investors optimal portfolio policy is *stationary* or *myopic* in the sense that the same optimal proportion is invested in each asset in every period. Sengupta (1989) has discussed these implications in some detail along with robustness issues.

## B. Structural Efficiency Under Uncertainty

The concept of structural efficiency developed by Farrell describes in a broad sense the degree to which an industry keeps up with the performance of its own best practice firms.

A concept of industrial efficiency related to structural efficiency was developed by Johansen (1972) who applied a linear programming (LP) approach to derive an industry production frontier from the stochastic input-output data of individual firms. He explicitly introduced the notions of statistical distribution of productive capacities among firms and a capacity utilization function in order to derive an aggregate industry-wide efficiency frontier. As in structural efficiency, one can compare two similar industries in respect of industrial efficiency by testing the distribution of units deviating from the full capacity level of utilization. The implications of such a formulation in stochastic DEA models are explored here. Such explorations would enhance the applicability of DEA models in two basic ways. One is the cross-section comparison of clusters of firms or plants, so that distortions induced by restrictions on the mobility of inputs can be evaluated. Second, the dynamic impact of the difference in capacity distribution and its utilization pattern may be analyzed in order to capture the technological progress embodied in such inputs as capital and durable equipment.

   A broad interpretation of Farrell's notion of structural efficiency is as follows: industry or cluster A is more efficient structurally than industry B, if the distribution of its best firms is more concentrated near its efficiency frontier for industry A than for B. Two difficulties arise very naturally with this definition. One is that the efficiency frontier may be probabilistic, when the input-output data are subject to a stochastic generating mechanism. Second, the distribution of the so-called best firms cannot altogether ignore the scatter of all other firms that are not necessarily the best. This is because the relative inefficiency of other firms serve to indicate the amount of potential gains which can be realized by reducing inefficiency.

   Consider a cluster of $N_1$ units each producing one output ($y_j$) with m inputs ($x_{ij}$) denoted by vector $x_j$ ($j=1,2,...,N_1$). Let $y^0$ be the maximum output feasible so that $y \leq y^0$. If $y^0 = f(x;\beta)$ is expressed in a linear form, the the efficiency frontier can be expressed in terms of the observed input-output vectors ($x_j, y_j; j=1,2,...,N_1$) as follows:

$$y_j^0 - u_j = \tilde{\beta}_0 + \sum_{i=1}^{m} \beta_i x_{ij} - u_j = y_j, \quad u_j \geq 0 \qquad (6.5.7)$$

where the non-negative errors $u_j$ indicate the efficiency gap. If the parameter $\theta = (\tilde{\beta}_0, \beta_1, ..., \beta_m)$ can be estimated from the observed input-output data ($x_j, y_j$), then he efficiency frontier $y^0$ can be determined and the statistical distribution of the obsered outputs $y_j$ around $y_j^0$ can be used as a measure of distribution of efficiency. However, the estimation problems for $\theta$ are complicated, since one of the basic assumptions of the maximum likelihood method of estimation is violated, unless the error $u_j$ follows a gamma distribution. To circumvent this problem we consider a tranformed version as

$$y_j = \beta_0 + \sum_{i=1}^{m} \beta_i x_{ij} + \varepsilon_j \qquad (6.5.8)$$

where $\beta_0 = \tilde{\beta}_0 - \mu, \varepsilon_j = \mu - u_j, \mu = E(u_j)$. Clearly this is of the least squares (LS) form and since the expected value $E(\varepsilon)$ of error $\varepsilon_j$ is zero we obtain

$E(y) = y^0 = \beta_0 + \sum_i \beta_i x_i$. From now on we assume that the observed input-output data satsify the LS form (6.5.8) where m is known to belong to a convex set $C(\mu) = \{\mu | a \leq \mu \leq b\}$. We have now a two-stage decision framework. In the first stage we estimate $\theta = (\beta_0, \beta_1, ..., \beta_m)$ conditional on a specific $\mu_0 \in C(\mu)$ by the LS model (6.5.8). At the second stage we vary $\mu^0$ in the set $C(\mu)$ till the maximum of the maximum likelihood estimates are reached in the first stage. Let $\mu^0$ be such an estimate and we assume it is unique and all the other estimates of $\theta_i$ conditional on $\mu^0$ are non-negative. In this framework we introduce the following definitions.

Definition 1
Cluster A is said to dominate B in the sense of structural efficiency (DSE), i.e. A DSE B if and only if $\overline{A} \geq \overline{B}$ and $\sigma_A \leq \sigma_B$ with at least one inequality strict. Here $\overline{A} = E(y_A), \overline{B} = E(y_B)$ and $\sigma_A^2 = var(y_A), \sigma_B^2 = var(y_B)$. Clearly this leads to the condition that the coefficient of variation of output in cluster A is less than that in B, i.e., $\sigma_A / \overline{A} < \sigma_B / \overline{B}$.

Definition 2
Cluster A dominates cluster B in the sense of first degree stochastic dominance (FSD), i.e. A FSD B if and only if the two cumulative distributions $F_A(t)$, $F_B(t)$ satisfy $F_A(t) \leq F_B(t)$ for all t and the inequality is strict for some $t = t_0$ in the relevant domain.

Definition 3
A cluster is said to be normalized if all the inputs $x_{ij}$ are divided by the observed outputs $y_j$, where the latter are assumed to be positive. The normalized cluster $\tilde{A}$ dominates another normalized cluster $\tilde{B}$ in the sense of first order aggregation, i.e. $\tilde{A}$ FAD $\tilde{B}$ if and only if it holds under aggregation of the two clusters that none of the efficient facets of $\tilde{A}$ turn out to be inefficient relative to $\tilde{B}$ and at least one efficient facet in $\tilde{B}$ turns out to be inefficient relative to $\tilde{A}$.

     Some comments on these definitions are in order. First of all, the dominance in structural efficiency (DSE) is essentially based on the mean variance characteristics of the output distribution in the two clusters. Hence if $\overline{A} = \overline{B}$ but $\sigma_A < \sigma_B$ we have A DSE B, i.e. the output variance in A is less than that in B. Thus the mean preserving scatter of output has less variance in A. Likewise if $\sigma_A = \sigma_B$, then DSE requires $\overline{A} > \overline{B}$ implying that the average level of output and therefore the efficient output is higher in cluster A. Note that if we have the additional information that the outputs in two clusters are normally distributed, then one can evaluate the probability $\beta$ that $y_A$ exceeds $y_B$

$$\text{Prob}\,[y_A > y_B] = F\left[\frac{\overline{y}_A - \overline{y}_B}{\sigma}\right] = \beta$$

that is

$$\overline{y}_A = \overline{y}_B + q\sigma, \quad \sigma^2 = \text{var}(y_A - y_B), \quad q = F^{-1}(\beta) \qquad (6.5.9)$$

Clearly if $\beta > 0.5$ then q is positive and hence $\overline{y}_A \geq \overline{y}_B$ for all $\beta$ in the interval $0.5 \leq \beta \leq 1$. Secondly, one has to note that SE depends on the form of the distribution of the error term $\varepsilon_j$ in (6.5.8) and also on the form $y^0 = f(x;\beta)$ of the efficiency frontier. Some empirical studies of local social insurance offices in Sweden during 1974-1984 by Bjurek et al (1990) showed that the distribution of input saving efficiency is substantially different for the quadratic, the Cobb-Douglas and the DEA models. They computed structural efficiency as a measure of technical efficiency for the whole industry (i.e. sector) by simply constructing a representative average office for the sector and regarding this office as any other office and then computing the individual efficiency measures for this office.

The stochastic dominance property is useful when the cumulative output distributions for the two clusters can be estimated. If one distributoin lies below the other, i.e. $F_A(t) < F_B(t)$ for all t, then A has FSD over B. Similarly stochastic dominance of second and higher orders can be defined. However in most practical situations the two output distributions would intersect, in which case FSD cannot hold. For such situations we may consider stochastic dominance under truncation. For example if the two distributions $F_A(t)$, $F_B(t)$ intersect once at c, we may consider the truncated variables $\hat{y}_A = \{y_A \mid y_A \geq c\}$ and $\hat{y}_B = \{y_B \mid y_B \geq c\}$ and ask if the two truncated distributions $F_{\hat{y}_A}(t), F_{\hat{y}_B}(t)$ satisfy for all $t \in (c, \infty)$, $c > 0$ the inequality

$$F_{\hat{y}_A}(t) \leq F_{\hat{y}_B}(t) \qquad (6.5.10)$$

with inequality holding strictly for some t. If it does, then $\hat{y}_A$ has the first degree stochastic dominance over the second truncated variable $\hat{y}_B$. Some empirical estimates of the efficiency distribution reported by Swedish social insurance offices by Bjurek et al (1988) showed that for some years the efficiency distributions for DEA and Cobb-Douglas models did not intersect at all, though for other years the two intersected.

Finally, the comparison of two normalized clusters $\tilde{A}$ and $\tilde{B}$ basically provides a measure of sensitivity of the efficiency frontier in $\tilde{A}$, when the normalized input-output data for cluster $\tilde{B}$ are added on. If the efficiency isoquant in $\tilde{A}$ is not altered by the addition of data points, then the efficiency frontier in $\tilde{A}$ is robust against $\tilde{B}$. However if the efficiency isoquant in $\tilde{A}$ has piecwise facets, and in at least one facet the efficient isoquant $\tilde{A}$ is below that of $\tilde{B}$, then the cluster $\tilde{A}$ has dominance over $\tilde{B}$ in the sense of FAD. Clearly if this dominance holds over all the facets, then the efficient isoquant $\tilde{A}$ is robust against $\tilde{B}$ in a global sense. Note that the dominance in the sense of FAD closely resembles the robustiness measure applied by Cubbin and Ganley (1987) to measure the efficiency of educational production in English local education authorities. Their measure computes the observed number of inefficient authorities for which the best practice authority forms the efficiency frontier. In tems of the DEA algorithm the most useful examples of best practice units are to be found only in heavily cited instances of best practice.

We have already emphasized the point that comparing two or more clusters in terms of their output distributions depends very critically on the specification of the efficiency frontier. Two applied methods are usually adopted for the purpose. One is the econometric method of estimating the frontier. Here only one frontier which may be nonlinear in form is specified or estimated for a given cluster and then the distribution of points analyzed. If the criterion of least absolute sum of errors is adopted, then this may lead to one LP omdel. The second method is the DEA algorithm by which N LP models are run for each unit in the cluster of N units. This specifies K distinct facets of efficiency ($K \leq N$), such that for the kth facet one can charecteize a distribution $F_k(\cdot)$ of points. Here one may have to compute N LP models to identify the K efficiency facets.

For example, in their empirical study Bjurek et al (1988) adopted the first approach to compute structural efficiency by simply constructing an average unit for the whole cluster and then estimating the individual measures of technical efficiency for this average unit. On the other hand, the DEA algorithm was recently applied by Cubbin and Ganley (1987) to identify units which are more often the best practice units, so that the latter can serve as targets for the relatively inefficient producers who may utilize the peer group identified by DEA as a guide to achieving their targeted improvements. We follow the method of mixtures for combining the individual distribution of output $F_k(d;\ \theta|y^0)$ for each facet $k=1,2,\ldots,K$ where $d = (X,Y)$ is the input-output data set. Hence the mixture model generating the data is of the form

$$F = \sum_{k=1}^{K} \pi_k F_k(d;\theta\,|\,y^0),\ \sum_{k=1}^{K} \pi_k = 1, \pi_k \geq 0 \qquad (6.5.11)$$

where $\pi_k$ is the proportion of the population in the kth group. Thus for any cluster we have the overall mixed distribution $F_A$ and also the component distributions $F_{kA}(\cdot)$ with weights $\pi_{kA}$. This idea of a distribution mixture with several component distributions is implicit in the empirical study by Farrell and Fieldhouse (1962) who noted bimodal frequency distributions for the efficiency measure. Clearly if k=2 and $F_1(\cdot), F_2(\cdot)$ have unequal modes then this would generate bimodal efficiency distributions.

We may now characterize the different concepts of efficiency when the data structure changes. For this purpose we introduce the definition of a mean preserving spread in data.

## Definition 4

The output distribution $F_{y_B}(t)$ in clusters B is said to be a mean preserving spread (MPS) or, a mean preserving increase in risk if and only if it is derived from $F_{y_A}(t)$ by taking weight from the center of the probability distribution and shifting it to the tails while keeping the mean of the distribution unchanged. This implies that output $y_B$ equals $y_A$ plus an uncorrelated noise term z such that it has a zero conditional mean $E(z|y_A)=0$ and a positive variance, i.e. $E(y_B) = E(y_A)$ and $\text{var}(y_B) = \text{var}(y_A) + \sigma^2$.

## Theorem 1

If the output data of cluster B has the MPS property with respect to cluster A, then cluster A dominates B in structural efficiency, i.e. A DSE B.

## Proof

By the definition of mean preserving spread we have $E(y_B) = E(y_A)$ and $\text{var}(y_B) > \text{var}(y_A)$. Hence $\sigma_A / \overline{A} < \sigma_B / \overline{B}$. Therefore A DSE B.

## Remark 1

The MPS property is sufficient but not necessary for DSE to hold.

## Remark 2

Under the MPS property the DSE relation in transitive, i.e. A DSE B and B DSE C implies that A DSE C.

## Theorem 2

Let the output cluster B be derived from A by truncation at a positive level c, i.e.

$$B = \hat{A} = \{A \mid y \geq c; c > 0\}$$

and assume that y is normally distributed. Then there exists a level $c^0$ of $c > 0$ such that $\hat{A}$ DSE A.

## Proof

Let $\mu(c)$, $\sigma^2(c)$ be the mean and variance of the truncated output variable belonging to cluster $\hat{A}$. Then by straightforward calculation

$$\mu(c) = \left[1 - F_A(\frac{c-\mu}{\sigma})\right]^{-1} \left[\mu + \sigma f_A(\frac{c-\mu}{\sigma})\right]$$

$$\sigma^2(c) = (\mu^2 + \sigma^2) + \left[1 - F_A(\frac{c-\mu}{\sigma})\right]^{-1}$$

$$\times f(\frac{c-\mu}{\sigma})[2\mu\sigma + c\sigma - \mu] - \mu^2(c)$$

where $\mu$ and $\sigma^2$ are the two parameters of the output distribution in A and $F(\cdot)$ is the cumulative distribution of a unit normal variate with density $f(\cdot)$. Since $F(t)$ and $f(t)$ are positive fractions for any finite $t$, we obtain $\mu(c) > \mu$ for a positive c. Also the variance $\sigma^2(c)$ of the truncated cluster $\hat{A}$ is a monotonically declining function for a positive c as Karlin (1982) has shown. Hence there exists a value $c^0$ such that $\sigma^2(c) \leq \sigma^2$ for all $c \geq c^0$. Hence $\sigma(c)/\mu(c) \leq \sigma/\mu$.

Remark 3
Karlin's theorem is more general in that the variance $\sigma^2(c)$ is a strictly decreasing monotonic function of the truncation level c as c increases in the positive domian for a whole class of log-concave density functions which include the normal, the gamma, the double exponential and the Polya density funciton.

Remark 4
If the output density in cluster A belongs to the log-convex class, e.g. log normal, then the effect of truncation is to increase the variance $\sigma^2(c)$ monotonically as c increases in the positive domain. Hence the DSE property of $\hat{A}$ over A may not hold in this case.

Theorem 3
If the two output distributions $F_{y_A}(t), F_{y_B}(t)$ intersect once at $t = c$, $c > 0$ so that $F_{y_B}(\tau) \geq F_{y_A}(\tau)$ for all $\tau \geq c$ and the inequality is strict for some $\tau \in (c, \infty)$, then $\hat{y}_A$ FSD $\hat{y}_B$ where $\hat{y}_A, \hat{y}_B$ are defined in (6.5.10).

Proof
Since the two ouput distributions have the single crossing porperty at $t = c$, $c > 0$ and the distribution of $y_A$ lies on or below that of $y_B$ for all $\tau \in (c, \infty)$, hence $\hat{y}_B$ cannot have first degree stochastic dominance over $\hat{y}_A$. Hence the result.

Remark 5

If the two density functions $\hat{f}_A$ and $\hat{f}_B$ for the non-negative random variables $\hat{y}_A$ and $\hat{y}_B$ have a common mean and finite variances then the variance of $\hat{y}_A$ is smaller than the variance of $\hat{y}_B$. This result holds because we have

$$\int_c^h F_{y_B}(\tau)d\tau \geq \int_c^h F_{y_A}(\tau)d\tau, \quad \text{for all h}$$

the property of second order stochastic dominance (SSD) satisfied here.

Remark 6
To test FSD of $\hat{y}_A$ over $\hat{y}_B$ over discrete sample observations i=1,2,...,N, we may use the statistic $D_k$

$$D_k = \begin{cases} \sum_{i=1}^k [\hat{f}_B(i) - \hat{f}_A(i)] \geq 0 & \text{for all } k \in [1,2,...,N] \\ \\ >0 & \text{for at least one } k \in [1,2,..,N) \end{cases}$$

where both $\hat{f}_B(i), \hat{f}_A(i)$ cannot be zero by assumption.

Theorem 4
If the normalized cluster $\tilde{A}$ has FAD (first order aggregation dominance) over another comparable cluster $\tilde{B}$, then there must exist at least one facet $EI_j(\tilde{A})$ in the efficient isoquant of $\tilde{A}$ which is on or below that of $EI_j(\tilde{B})$. Hence the existence of FAD can be tested over discrete observations by the decision rule $(DR_k)$:

$$DR_k = \begin{cases} \sum_{i=1}^k [EI_j(\tilde{B}) - EI_j(\tilde{A})] \geq 0, & \text{for all } k \in K \\ \\ >0, & \text{for at least one } k \in K \end{cases}$$

where $K = (1,2,...,h)$ is a finite set of comparable facets. Furthermore the probability $p_k$ of the event $DR_k$ must be positive whenever FAD holds.

Proof
By definition FAD implies that the efficient isoquant of cluster $\tilde{B}$ cannot lie below that of $\tilde{A}$. If the efficient isoquant in each cluster has a set of K facets, this implies that in at least one facet the efficient isoquant $\tilde{A}$ must lie on or below

that of $\tilde{B}$. Hence the decision rule $DR_k$. By the same reasoning the probability of the event $\{DR_k \geq 0$ for all $k \in K$ and $DR_k > 0$ for at least one $k \in K\}$ cannot be zero. Hence the result.

Remark 7

Whereas stochastic dominance and dominance in structural efficiency deal with the distribution of output in the two clusters, the dominance under aggregation basically tests the distance between the two efficient isoquants or their corresponding facets. Hence it is easier to apply, though we have to note that it depends on the way the efficient isoquant is specified in each cluster parametrically or non-parametrically.

Remark 8

If the efficient isoquants for the two clusters intersect once, one could again compare the two truncated but efficient isoquants to the right (or left) of the intersection point.

We now consider the problem of measuring industrial efficiency by the DEA approach.

In Johansen's approach a cluster is viewed as an industry comprising N production units with different production capacities and different input coefficients. The output $y_j$ of unit $j$ is viewed separately from the capacity output $\bar{y}_j$ where $j \in I_N = \{1, 2, ..., N\}$, so that the input coefficients $\bar{a}_{ij} = \bar{x}_{ij} / \bar{y}_j$ are taken at the full capacity level of utilization, i.e., $\bar{x}_{ij}$ is the input type i required at the full capacity level of output. The production frontier is then obtained by maximizing the total industry output. The resulting LP model is

$$\max y_T = \sum_{j=1}^{N} y_j$$

such that

$$\sum_{j=1}^{N} \bar{a}_{ij} y_j \leq x_i, i \in I_m \qquad (6.5.12)$$

$$0 \leq y_j \leq \bar{y}_j, j \in I_N$$

Hence $x_i = \sum_{j=1}^{N} x_{ij}$ is the aggregate input of type i. Clearly the LP model has a finite optimal solution if the aggregate inputs $(x_i)$ and the input coefficients $a_{ij}$ are finite and positive. Two interpretations are given to the objective function: either as a normative model for the industry planner, or as positive theory supported by the competitive market behavior under decentralized decision-making.

Note two basic differences of this model from the structural efficiency approach of Farrell. One is that the input coefficients $\bar{a}_{ij} = \bar{x}_{ij}/\bar{y}_j$ here are appropriate only to the full capacity level of utilization, which is different from the observed input-output ratios $a_{ij} = x_{ij}/y_j$ if $x_{ij} \neq \bar{x}_{ij}$ and $y_j \neq \bar{y}_j$. Secondly, the distribution of the capacity outputs $\bar{y}_j$ over firms denoted by $F(\bar{y})$ is to be distinguished here from that of observed outputs denoted by $G(y)$ say. One could thus characterize several facets of efficiency in this model. First of all, if it were feasible to secure $y_j^0 = \bar{y}_j$ (where $y^0$ denotes the optimal solution) for all $j \in I_N$ such that $\sum_j \bar{a}_{ij} y_j = x_i, i \in I_m$ then it would represent full capacity utilization of all available inputs and outputs. This is identical with the neoclassical model of full employment level of output which allows full capacity utilization of available resources. Secondly, if for at least one j it holds that $y_j^0 < \bar{y}_j$, or for one i it holds that $\sum_j \bar{a}_{ij} y_j^0 < x_i$ we would have inefficiency due to unutilized capacity, or unemployed resources (i.e., $x_i^0 = \sum_j \bar{a}_{ij} y_j^0 < x_i$). The two distributions $F(\bar{y})$ and $G(y^0)$ would no longer be identical. Thirdly, assume that the capacity output levels are not available and we have (6.5.12) in the form $y_j \geq 0$. Furthermore the observed input coefficient $a_{ij}$ which replace $\bar{a}_{ij}$ are given. Then the optimal vector $y^0 = (y_j^0)$ provides a characterization of efficiency, since by duality one can write

$$\min_\beta \beta' x \quad \text{such that } \beta'A \geq e', \beta \geq 0 \tag{6.5.13}$$

where the vector matrix notation is used with $x = (x_i)$, $A = (a_{ij})$, $e = (e_i)$, $e_i = 1$ and $\beta = (\beta_i)$. Let $\beta^0 = (\beta_i^0)$ be the optimal solution vector and $z^0 = (z_j)$ be the corresponding imputed outputs $z_j^0 = \sum_j \beta_i^0 x_{ij}$. If $z_j^0 > y_j$ we have output inefficiency, since the observed output is less than its efficient level given by the imputed quantity $z_j^0$. The distance of the distribution $G(y)$ of observed $y_j$ values from that $F(z^0)$ of the efficient imputed outputs would provide a characterization of inefficiency.

Another variant of the dual LP model (6.5.13) is considered by the DEA model, which transforms the single LP model (6.5.13) into N LP models each having the identical constraint set but with a separate objective function, e.g.

$$\min_{\beta(k)} g_k = \beta' x_k \quad \text{such that } \beta'X \geq y', \beta \geq 0 \tag{6.5.14}$$

Denote the optimal solution vector by $\beta^0 = \beta^0(k)$ and vary $k \in I_N$ to generate N optimal solution vectors. Once again for reach fixed k one could compute the distribution $F_k = F(z^0(k))$ of $z^0(k) = \beta^{0'}(k)X$ and compare it with the distribution G(y). A suitable statistical distance function measuring the closeness of the two distributions, e.g. Kullback-Leibler information number $I(F_k, G)$

$$I(F_k, G) = \int \left[ \ln\left( \frac{f_k(t)}{g(t)} \right) \right] f_k(t) dt \qquad (6.5.15)$$

where $f_k(\cdot), g(\cdot)$ are the respective density functions, provides such a distance function which can be statistically tested by the Akaike criterion when the distributions $F_k(\cdot)$ and $G(\cdot)$ are estimated by their empirical sample counterparts.

A slightly different variant of the LP model (6.5.14) is obtained when we replace objective function $g_k$ by its sum, i.e.

$$\min_{\beta} g = \sum_{k=1}^{N} g_k = N\beta'\overline{x}, \quad \overline{x} = \frac{\Sigma x_k}{N} \qquad (6.5.16)$$

In this case we have one LP model to determine the efficiency frontier. Denoting the optimal solution by vector $\overline{\beta}^0$ one could again compute the distribution $F(\overline{z}^0)$ of efficient output $\overline{z}^0$ and compare it with G(y), the distribution of the observed output.

Note that for the set of (N+1) LP models (6.5.14) and (6.5.16) there exists at least one LP for which the distance function (6.5.15) is the minimum, i.e.

$$I(F_k, G) \geq I(F_h, G), \quad k = 1, 2, ..., N + 1$$

This distribution $F_h(\cdot)$ may then be interpreted as the best approximation of the efficiency frontier by the observed data G since it maximizes the entropy which is the negative of $I(F_k, G)$.

There is an alternative way of looking at the distance between G and the various optimal output distributions $F(z^0(k))$, $F(\overline{z}^0)$ and $F(\overline{y})$. If we consider $F(\overline{y})$ as the true distribution of efficient output, which of the distributions $\{F(z^0(k)), k=1,...,N+1\}$ comes closest to the true distribution? By the maximum entropy principle this will be given by $F_h$ satisfying the following condition

$$I(F_k, F(\overline{y})) \geq I(F_h, F(\overline{y}))$$

One could apply this criterion to empirical distribution functions derived from the various LP models above. Some empirical applications of this approach have been discussed by Sengupta (1990, 1992)

Two types of extensions may be suggested here. One is the case of multiple outputs where multivariate output distributions are to be compared in a nonparametric manner through the stochastic dominance approach. Secondly, the time series changes in efficiency and its distributions may need to be analyzed. This is particularly important in industries where seasonal and cyclical fluctuations are present, e.g., agriculture.

# Bibliography

Arrow, K., The Economic Implications of Learning by Doing, *Review of Economic Studies*, 29, 155-174, 1962.

Artus, P. and Muet, P. *Investment and Factor Demand*, North Holland, Amsterdam, 1990.

Athanassopoulos, A. and Thanassoulis, E, Separating Market Efficiency from Profitability and its Implicatins for Planning, *Journal of the Operational Research Society*, 46, 20-34, 1995.

Battese, G.E. and Coelli, T.J., Frontier Production Functions, Technical Efficiency and Panel Data with Application to Paddy Farmers in India, *Journal of Productivity Analysis*, 3, 153-169, 1992.

Battese, G.E. and Coelli, T.J., A Model for Technical Inefficiency Effects in a Stochastic Frontier, *Empirical Economics*, 20, 325-332, 1995.

Baumol, W.J., Panzar, J.C. and Willig, R.D., *Contestable Markets and the Theory of Industry Structure*, Harcourt Brace, New York, 1982.

Benasy, J., *The Economics of Market Disequlibrium*, Academic Press, New York, 1982.

Bjurek, H., Hjalmarsson, L. and Forsund, F.R., Deterministic Parametric and Nonparametric Estimation of Efficiency in Service Production: A Comparison, *Journal of Econometrics*, 46, 213-228, 1990.

Bogetoft, P., Incentive Efficient Production Frontiers: An Agency Perspective on DEA, *Management Science*, 40, 959-968, 1994.

Byrnes, P., Fare, R. and Grosskopf, S., Measuring Production Efficiency: An Application to Illinois Stripmines, *Management Science*, 30, 671-681, 1984.

Carroll, R. and Ruppert, D. *Transformation and Weighting in Regression*, Chapman and Hall, London, 1988.

Charnes, A., Cooper, W.W. and Rhodes, E., Measuring the Efficiency of Decision-making Units, *European Journal of Operational Research*, 2, 429-444, 1978.

Cooper, W.W. and Gallegos, A., A Combined DEA-Stochastic Frontier Approach to Latin American Airline Efficiency Evaluations, Working Paper, University of Texas, Austin, 1992.

Cornwell, C. and Schmidt, P., Production Frontiers and Efficiency Measurement, in Matyas, L. and Sevestre, P. eds., *The Econometrics of Panel Data*, Kluwer, Boston, 1995.

Dalen, D.M., Testing for Technical Change with Data Envelopment Analysis, Working Paper No. 16, Department of Economics, University of Oslo, 1993.

Dorroh, J.R., Gulledge, T.R. and Wormer, N.K., Investment in Knowledge: A Generalization of Learning by Experience, *Management Science*, 40, 947-958, 1994.

Farrell, M.J. and Fieldhouse, M., Estimating Efficiency in Production Function Under Increasing Returns to Scale, *Journal of Royal Statistical Society*, Series A, 125, 252-267, 1962.

Fine, C.H. and Freund, R.M., Economic Analysis of Production Flexible Manufacturing Systems' Investment Decisions, *Proceedings of the Second ORSA-TIMS Conference on Flexible Manufacturing Systems*, North Holland, Amsterdam, 1986.

Fine, C.H., Quality Improvement and Learning in Productive Systems, *Management Science*, 32, 1301-1315, 1986.

Foss, M.F., The Utilization of Capital Equipment: Postwar Compared with Prewar, *Survey of Current Business*, 43, 8-16, 1963.

Ganley, J.A. and Cubbin, J.S., *Public Sector Efficiency Measurement: Applications of Data Envelopment Analysis*, North Holland, Amsterdam, 1992.

Greene, W.H., A Gamma Distributed Stochastic Frontier Model, *Journal of Econometrics*, 46, 141-164, 1990.

Greenwood, J., Hercowitz, A. and Krussel, P., Long-run Implications of Investment-specific Technological Change, *American Economic Review*, 87, 342-359, 1997.

Grossman, S. and Stiglitz, J., On the Impossibility of Informationally Efficient Markets, *American Economic Review*, 70, 393-408, 1980.

Hall, R.E., Invariance Properties of Solow's Productivity Residual, in Diamond, P. (ed.), *Growth, Productivity and Unemployment*, MIT Press, Cambridge, Massachusetts, 1990.

Hardle, W. and Bowman, A.W., Bootstrapping in Nonparametric Regression, *Journal of American Statistical Association*, 83, 123-127, 1988.

Henderson, J.M., Efficiency and Pricing in the Coal Industry, *Review of Economics and Statistics*, 38, 50-60, 1956.

Hodrick, R. and Prescott, E.C., Post-war U.S. Business Cycles: An Empirical Investigation, Working Paper, Carnegie-Mellon University, 1993.

Holt, C.C., Modigliani, F. Muth, J.K. and Simon, H.A., *Planning Production Inventory and Workforce*, Prentice Hall, Englewood Cliffs, New Jersey, 1960.

Hjalmarsson, K., Kumbhakar, S. and Heshmati, A., DEA, DFA and SFA: A Comparison, *Journal of Productivity Analysis*, 7, 303-327, 1996.

Hulten, C.R., Growth Accounting when Technical Change is Embodied in Capital, *American Economic Review*, 82, 964-980, 1992.

Ippolito, R.A., Efficiency with Costly Information: A Study of Mutual Fund Performance, 1965-84, *Quarterly Journal of Economics*, 104, 1-23, 1989.

Johansen, L., *Production Functions*, North Holland, Amsterdam, 1972.

Johnson, N.L. and Kotz, S., *Distributions in Statistics*, Houghton Hifflin, New York, 1970.

Jovanovic, B., Learning and Growth, in Kreps, D.M. and Wallis, K.F., *Advances in Economics and Econometrics*, Cambridge University Press, New York, 1997.

Karlin, S., Some Results on the Optimal Partitioning of Variance and Monotonicity with the Truncation Level, *Statistics and Probability*, North Holland, Amsterdam, 1982.

Kennan, J., The Estimation of Partial Adjustment Models with Rational Expectations, *Econometrica*, 47, 1441-1456, 1979.

Kennedy, J.E., An Analysis of Time Series Estimates of Capacity Utilization, Working Paper, Federal Reserve Board, Washington, D.C., 1995.

Klein, L.R. and Long, V., Capacity Utilization: Concept, Measurement and Recent Estimates, *Brookings Papers on Economic Activity*, 3, 743-761, 1973.

Kumar, C.K. and Sinha, B.K., Efficiency Based Decision Rules for Production Planning and Control, *International Journal of Systems Science*, 29, 1265-1280, 1998.

Leibenstein, H., Allocative Efficiency versus X-efficiency, *American Economic Review*, 66, 392-415, 1966.

Lothgren, M. and Tambour, M., Testing Scale Efficiency in DEA Models: A Bootstrapping Approach, Working Paper, Stockholm School of Economics, 1997.

Lucas, R.E., Making a Miracle, *Econometrica*, 61, 251-272, 1993.

Maddala, G.S., Limited Dependent and Qualitative Variables in Economics, Cambridge University Press, New York, 1983.

Morrison, C. and Berndt, E., Short-run Labor Productivity in a Dynamic Model, *Journal of Econometrics*, 16, 339-365, 1981.

Murthi, B., Choi, Y. and Desai, P., Efficiency of Mutual Funds and Portfolio Performance Measurement: A Nonparametric Measurement, *European Journal of Operational Research*, 98, 408-418, 1997.

Norsworthy, J.R. and Jang. S.L., *Empirical Measurement and Analysis of Productivity and Technological Change*, North Holland, Amsterdam, 1992.

Oum, T. and Yu, C., *Winning Airlines: Productivity and Cost Competitiveness of the World's Major Airlines*, Kluwer, Dordrecht, 1998.

Pulley, L.B. and Braunstein, Y.M., A Composite Cost Function for Multiproduct Firms with an Application to Economies of Scope in Banking, *Review of Economics and Statistics*, 74, 221-230, 1992.

Reifschneider, D. and Stevenson, R., Systematic Departures from the Frontier: A Framework for Analysis of Farm Inefficiency, *International Economic Review*, 32, 715-723, 1991.

Romer, P.M., Are Nonconvexities Important for Understanding Growth, *American Economic Review*, 80, 97-103, 1990.

Selten, R., Elementary Theory of Slack-ridden Imperfect Competition, in J.E. Stiglitz and G. Mathewson, eds., *New Developments in the Analysis of Market Structure*, MIT Press, Cambridge, 1986.

Sengupta, J.K., Structural Efficiency in Stochastic Models of Data Envelopment Analysis, *International Journal of Systems Science*, 21, 1047-1056, 1990.

Sengupta, J.K., On the Price and Structural Efficiency in Farrell's Model, *Bulletin of Economic Research*, 44, 281-300, 1992.

Sengupta, J.K., *Econometrics of Information and Efficiency*, Dordrecht: Kluwer Academic Publishers, 1993.

Sengupta, J.K., Measuring Dynamic Efficiency Under Risk Aversion, *European Journal of Operational Research*, 74, 61-69, 1994.

Sengupta, J.K., *Dynamics of Data Envelopment Analysis*, Kluwer, Boston, 1995a.

Sengupta, J.K., Dynamic Farrell Efficiency: A Time Series Application, *Applied Economics Letters*, 2, 363-366, 1995b.

Sengupta, J.K., Data Envelopment Analysis: A New Tool for Managerial Efficiency, *International Journal of Systems Science*, 27, 1205-1210, 1996a.

Sengupta, J.K., Adjustment Costs in Mean Variance Efficiency Analysis, *International Journal of Systems Science*, 27, 551-559, 1996b.

Sengupta, J.K., Recent Models in Data Envelopment Analysis: Theory and Applications, *Applied Stochastic Models and Data Analysis*, 12, 1-26, 1996c.

Sengupta, J.K., Systematic Measures of Dynamic Farrell Efficiency, *Applied Economics Letters*, 3, 91-94, 1996d.

Sengupta, J.K., The Efficiency Distribution Approach in Data Envelopment Analysis, *Journal of the Operational Research Society*, 47, 1387-1397, 1996e.

Sengupta, J.K., New Efficiency Theory: Extensions and New Applications of Data Envelopment Analysis, *International Journal of Systems Science*, 29, 255-265, 1998a.

Sengupta, J.K., Testing for Technical Change: A Nonparametric Application, *Applied Economics Letters*, 5, 43-46, 1998b.

Sengupta, J.K., A Dynamic View of the Portfolio Efficiency Frontier, *Computers and Mathematical Applications*, 18, 565-580, 1989a.

Sengupta, J.K., *Efficiency Analysis by Production Frontiers: The Nonparametric Approach*, Kluwer, Boston, 1989b.

Sengupta, J.K., The Measurement of Dynamic Productive Efficiency, *Bulletin of Economic Research*, 51, 111-124, 1999a.

Sengupta, J.K., Farrell-type Efficiency Under Demand and Price Fluctuations, *Applied Economics Letters*, 6, 121-125, 1999b.

Schefczyk, M., Operational Performance of Airlines: An Extension of Traditional Measurement Paradigms, *Strategic Management Journal*, 14, 301-317, 1993.

Shang, J. and Sueyoshi, T., A Unified Framework for the Selection of a Flexible Manufacturing System, *European Journal of Operational Research*, 85, 297-315, 1995.

Simar, L. and Wilson, P.W., Sensitivity Analysis of Efficiency Scores: How to Bootstrap in Nonparametric Frontier Models, *Management Science*, 44, 49-61, 1998.

Solow, R.M., A Contribution to the Theory of Economic Growth, *Quarterly Journal of Economics*, 70, 65-94, 1956.

Spence, A.M., The Learning Curve and Competition, *Bell Journal of Economics*, 12, 49-70, 1981.

Sueyoshi, T., Measuring Efficiencies and Returns to Scale of Nippon Telegraph and Telephone in Production and Cost Analyses, *Management Science*, 43, 779-796, 1997.

Sueyoshi, T., DEA Window Malmquist Analysis and Kruskal-Wallis Test, Working Paper, Science University of Tokyo, 1999.

Tofallis, C., Input Efficiency Profiling: An Application to Airlines, *Computers and Operations Research*, 24, 253-258, 1997.

Tulkens, H. and Eeckut, P., Nonparametric Efficiency: Progress and Regress Measures for Panel Data, *European Journal of Operational Research*, 80, 474-499, 195.

Winston, G.C., The Theory of Capacity Utilization and Idleness, *Journal of Economic Literature*, 12, 125-134, 1974.

Womer, N.K., Learning Curves, Production Ratio and Program Costs, *Management Science*, 25, 312-319, 1979.

# Index